# *ANCESTRY QUEST*

# *ANCESTRY QUEST*

### HOW STORIES FROM THE PAST
### CAN HEAL THE FUTURE

## MARY BETH SAMMONS

Published in the United States by Viva Editions, an imprint of Start Midnight, LLC, 221 River Street, 9th Floor, Hoboken, New Jersey 07030.

Printed in the United States.

Cover design: Jennifer Do
Cover photograph: Shutterstock
Text design: Frank Wiedemann

First Edition.

10 9 8 7 6 5 4 3 2 1
Trade paper ISBN: 978-1-63228-069-5
E-book ISBN: 978-1-63228-125-8

*For my mother Isabel,*
*I will keep my promise to find him.*
*For my grandfather Austin, why?*

# TABLE OF CONTENTS

## INTRODUCTION

What makes us who we are? What combination of history, genes, ancestry, experience, and that intangible thing called the spirit defines us? Commercial DNA testing and websites like 23andMe.com, Ancestry.com, FamilySearch.org, FindMyPast.com, and MyHeritage.com, along with TV shows including NBC's *Who Do You Think You Are?*, PBS's *Finding Your Roots* and *Genealogy Roadshow*, and CNN's *Roots: Our Journeys Home*, have exploded in popularity and offer profound answers for the curious. Where did my ancestors come from? To whom am I biologically related? Who am I?

Internationally, genealogy searches have become one of the world's most popular pastimes.

As Frank Delaney wrote in *Shannon*, "Within our origins we search for our anchors, our steadiness. And everyone's journey to the past is different. It might be found in a legend or in the lore of an ancestor's courage or an inherited flair. Or it might be found simply by standing on the earth once owned by the namesake tribe, touching the stone they carved, finding their spoor. In all cases we are drawn to the places whence they came—because to grasp who they were may guide what we might become."[1]

Historically these stories were told by elders, often by the fireside. In modern times, we follow the footsteps of our ancestors through internet research, church records, photo albums, journals, recipe boxes, interviews with living family members, and spitting into at-home DNA test kits.

For many of us, this can result in dramatically inconsistent findings compared to our understanding of who we are. "Looking at your genetic data might uncover information that some people find surprising. This information can be relatively benign. At other times, the information you learn can have profound implications for both you and your family," warns the boilerplate legal language at 23andMe.com, the DNA testing site.[2]

The four largest companies offering home DNA testing kits are AncestryDNA, MyHeritage, FamilyTreeDNA, and 23andMe. As of late 2019, over 28.5 million people had bought a genetic testing kit from one of those services, according to an article in *Kiplinger*.[3] By the estimates of Global Market Insights, the global market for direct-to-consumer genetic testing will surpass $2.5 billion by 2024.[4]

The popularity of these at-home DNA testing kits is skyrocketing. "By the start of 2019, more than 26 million consumers had added their DNA to four leading commercial ancestry and health databases," according to estimates by the *MIT Technology Review*.[5]

Likewise, "About one-in-seven U.S. adults (15%) say they have ever used a mail-in DNA testing service from a company such as AncestryDNA or 23andMe," according to a Pew Research Center survey that was published in 2019.[6]

Most seekers (87 percent) said they were eager to learn about their family origins, and a notable share said the results surprised them, according to the Pew study. About four in ten of those who used mail-in DNA testing said they were surprised by results about where their ancestors came from. Additionally, 27 percent said they were surprised by the ethnic background of their ancestors, 26 percent about health or family medical history, and 27 percent (more than a quarter) said they learned about a relative they did not know about.[7]

That speaks volumes about the power of a simple DNA test and ancestral sleuthing to upend an identity and dismantle a family story.

Just ask the growing ranks of seekers who have received life-changing results. For many, this process has recast entire lives with surprises including shocking lineages, long-lost siblings, and family secrets that might have been buried for decades. For many, it has opened questions about heritage, ethnicity, race, culture, and privacy.

With more than a fourth of people making surprising discoveries, DNA support groups are popping up on Facebook. Another project, NPE Friends Fellowship, offers dozens of highly specific support groups for various relatives, including those who uncovered and those who kept a family secret. The nonprofit organization also hosts symposia, conferences, and even cruises for people dealing with an "NPE" ("Not Parent Expected") discovery. In 2019, 23andMe.com launched a resource page exclusively for users who have discovered unexpected results through their tests.

For others, delving into family stories can lead to self-discovery and a broader sense of connection. Some say it can be healing. "Research on family history argues it performs the task of anchoring a sense of 'self' through tracing ancestral connection and cultural belonging, seeing it as a form of storied 'identity work,'" according to a study by the University of Manchester's Wendy Bottero.[8]

This book includes the stories of people who have made decisions that could alter their futures—or really, what they think they know about their pasts. It might start as a just-for-fun DNA test, but ultimately those who are spitting in tubes and sending them off in the mail or are burning the midnight oil hunting through ancestry records on the internet share a common goal: they all want to know what they are made of.

*Ancestry Quest: How Stories of the Past Can Heal the Future* shares the remarkable journeys of these real-life persons who delved deeper into their family mysteries. Many have uncovered, quite by accident, that their family isn't entirely what they thought. These moving and thoughtful stories will take you on a roller coaster of emotions and will resonate with all who are exploring their histories. You know who you are and what the questions are: "Who am I?" and "Why am I who I am?" The journeys are dramatically different—sad and happy all at once, promising to redefine family in a way that is more honest and relevant. These heartwarming and wrenching, intimate and inspiring stories are about the lessons learned along the way in search of the truth.

It seems everywhere we turn these days, someone is talking about someone whose coworker or cashier at the coffee shop just found out some kind of shocking discovery about the family they thought they had.

I know firsthand. This book was largely inspired by my own ancestral quest.

As a veteran newspaper journalist and author, I was accustomed to digging for information to tell other people's stories. But several years ago, the tables turned. On the morning my eighty-six-year-old mother was diagnosed with a rare duodenum cancer and given four months to live, she reached out to me with one urgent request: "Mary, please find out whatever happened to my father."

Through this book, I share that journey and the remarkable, life-changing stories of real-life people who yearn to know more about their ancestors. I believe these personal odysseys are universal stories, ones that I hope will captivate and inspire you to begin your search.

From my conversations with the ancestry seekers featured here and with my friends, colleagues, and my own family, I believe our family stories have much to teach us for generations to come.

"When we tell the secret that we feel sets us so completely apart from everyone else, we discover that it doesn't and that to connect with others is valuable and powerful," says Dani Shapiro, author of the book *Inheritance: A Memoir of Genealogy, Paternity, and Love.*[9]

In our ever-changing world, there is no substitute for a legacy of family stories to provide direction, a sense of identity, and to serve as a reminder of the grit and inspiration needed to move ahead. At best, I hope this book will help us look with compassion at the lives and circumstances of the ones who came before us and to realize how they have helped shape our own lives.

This is especially timely in 2020, the year of the first

Census in ten years. When the data is released, genealogy tests can help us better understand our place among our peers in America.

Join me on these journeys. I hope that they will inspire your own.

## CHAPTER ONE

## ORDINARY PEOPLE, AMAZING DISCOVERIES:

When DNA and ancestry stories change your life

Thanks to a national scout jamboree in Virginia and a volunteer project helping teens discover more about themselves through an ancestry search project, Edgie Donakey says he "fell in love with what technology can do to help families discover their stories."

"Seeing these young people on their computers, finding out who they are by learning more about the relatives that came before them, opened my eyes to how powerful these discoveries really are," says Donakey.

He's also personally experienced the poignant moments when an ancestry connection is suddenly made.

One of those moments happened when he was in downstate Illinois, where he had traced some of the Donakey family to a farm they ran. When he was visiting a small historical society there, he met "two little old ladies who were the sweetest things," says Donakey. The Donakeys had long since left the area and moved to the Northwest, but the women remembered the mark the family made in the town.

"They were so excited to find out I was a member of the Donakey family," he says. "And I was like a seven-year-old on Christmas morning—it was that thrilling for me."

Donakey's experience speaks volumes about the magic that happens when people suddenly stumble on an ancestry discovery and make a connection that previously eluded them. Since that Eagle Scout camp sixteen years ago, he's been committed to helping people across the globe discover the stories of their ancestors.

As the vice president of strategic relations and deputy chief genealogy officer of FamilySearch International, Donakey monitors the worldwide landscape of genealogy, looking for the newest and best tools to put into the hands of future generations of ancestry seekers.

"What makes everyone so fascinated by searching their ancestry is that all of us have powerful stories of those who came before us and alongside us that we want to hand down to our children and our children's children," says

Donakey, who made that midcareer switch sixteen years ago to FamilySearch after his fateful Eagle Scout volunteer weekend. Prior to that, he had carved out a highly successful career in the fast-paced corporate world of mergers and acquisitions, designing and implementing strategies for Fortune 500 companies. "Whether you're a multimillionaire or on a fixed income, you can do all the searching you want on the ancestry sites. And those searches can result in some very exciting moments."

During his business travels around the world, he says he's always thrilled when he's at the airport and spots travelers on their laptops searching the ancestry sites.

"Everywhere I go, these conversations are striking up," he says.

Through his work, he and his colleagues have helped people find many surprises. Of course, sometimes digging into the family tree unearths pieces of a bigger story than the one you envisioned. As can digging through old suitcases and memorabilia buried in the back of closets and other places where ancestors hid their secrets.

Often, the answers aren't what we wanted to hear. If your experience is anything like some, you may find a dramatic "switched at birth" situation, a sibling you never knew about, or other results that are confounding.

Buyer beware. Even the commercial genetic tests spell out the possibility of such a bombshell.

And what happens when you aren't even exploring

your own ancestry, but out of the blue the phone rings and it's a long-lost, never-known relative who has tracked you down?

More often than expected, the findings can force you to reimagine your identity. For some the confusion leads to outright shock. For others, it confirms an inner knowing that something was just not right.

What we've come to learn in sharing the stories in this book is that knowing the truth, even if it feels harsh or hard to accept at first, can be healing. In some cases, it can give us a sense of empathy when we realize we are all human and sometimes we find ourselves making decisions that have a ripple effect for generations to come.

The stories of people in this chapter underscore the importance of sharing our stories as an act of growth, healing, and ultimately selflessness. When we find these authentic stories—sad, funny, hopeful, or tragic—we can begin to narrate our own.

"There are infinite stories in the records out there and infinite numbers of people seeking them," says Donakey. "It's important for all of us to find them."

### An Ancestry Adventure: When DNA testing unearths the family secret

*While trying out two of the most popular DNA testing services—Ancestry.com and 23andMe. com—all Carole Hines wanted to know was why her brother was so tall, so blond, and so strikingly opposite looking compared to her own five-foot-three, black-haired self.*

*The questions started at a young age, when she intuitively knew that "something made me really different from my brother and sister."*

*That simple knowing would take Hines, sixty-nine, who lives with her wife, Mavis, in both San Francisco and New York City, on a long journey of race and ethnicity.*

*It all started several years ago when her brother died suddenly in his sleep of a massive heart attack. He had recently tested his DNA, and his results showed he was 99 percent of European descent.*

*More than forty years before, when Hines first introduced her brother to Mavis for the first time, she remembers Mavis's shock. When the tall, blue-eyed, "very Viking-looking" man walked in the door, Mavis looked at her, and she could just feel her thinking, "What, this is your brother, really?"*

*Fast-forward to 2017, when Hines took her DNA test. She couldn't believe the results, so she took a second one with another company and discovered she was mostly Latino, with traces of Native American, Ashkenazi Jewish, and Basque.*

> *Ironically, coincidentally, or call it providentially, Hines, who is a frequent globetrotter, says every time she's in Spain (her favorite country) people approach her speaking Castilian Spanish and asking for directions.*
>
> *"It's funny, but I've always loved the food. If I could, I would live on cheese enchiladas," says Hines.*

Unlike many people who discover that their DNA is startlingly different from their family members', Hines said she didn't find the news troubling or opening any wounds. Instead it was more like she could better understand the reason for the scars she had carried with her for more than sixty years.

"I always felt like I was such an oddball, but now I know the truth," says Hines.

The truth, it turns out, is that her biological father, Joe, who was from Mexico, was quite the Casanova. During the 1950s, her mother, Joyce, worked in Los Angeles for the Teamsters. Hines learned through conversations with one of her half brothers that her biological father was a union organizer and spent a lot of time traveling to Los Angeles—and he "had a hobby of dating married women."

"My father and mother had their own issues, and he was probably not around, so along comes this man who jangles his money around, wines and dines her, and probably says, 'Oh honey, I'm going to get a divorce.' And then he vanishes."

Evidently that happened at least nine times to different women, as Hines has uncovered eight half brothers. "I'm the only female I know of."

Since she's connected with her biological father's heirs, she says most of them have embraced her and are now her Facebook friends. Many live in Southern California, and she has formed relationships with some of them and some nieces and nephews.

The oldest in her family, Hines is four years apart from her brother. Her brother and her sister are the biological children of the man she thought was her father and called "Daddy" her whole life. He had black hair and brown eyes, and she says she thought she looked like him, or wanted to believe she did. His family traces back to the *Mayflower*, and her mother, with her auburn hair and light skin, "was very Irish." Both of her parents are deceased.

The finding was freeing, says Hines.

Throughout her life, Hines says she remembers marching to a different drummer than her siblings and family members. Her intellectual bent and passion to crusade for people who can't speak up for themselves were mainstays.

"I couldn't tolerate prejudice," says Hines, who grew up in Southern California but in high school moved to Memphis, Tennessee, to live with her father, who had relocated there after her parents' divorce. "I am the biggest coward in the world, but one day when I was in class and someone said a very derogatory word against a classmate, I remember jumping out of my chair and saying, 'Don't call people that.'"

She also remembers a "not-so-happy" childhood after her parents divorced when she was eight.

"I worked hard at school and read a lot, but I always felt very anxious and vulnerable," she says. "I didn't know who I was. My job was to take care of the younger kids, so I just lost myself in books and school."

Today, she and Mavis are committed pacifists who, she says, "have participated in more peace marches than I can count."

Recently she posted a childhood photo of herself and her friends at a birthday on her Facebook page, which speaks to the strong connection she has made to her newly discovered Latino heritage and the relatives—eight half brothers and their families—she's discovered from the DNA matches.

"The picture is of me at my birthday, and there I am, this little Mexican girl with a full head of raven-black hair surrounded by a sea of white kids," she says. A husband of one of her biological family members saw the photo and asked his wife, "Is that you?" Hines said.

## LEGACY LESSON

"I've never understood racism or prejudice and anything that diminishes people because of their race or ethnicity or how they live. Now I better understand what I instinctively knew in my pores, that I was of a different color, that I was a little bit different. Maybe I feel so strongly because I was fighting unconsciously for myself."

### Caught by Surprise: A relative's DNA search reveals secret family members

*Bridget (Wingert) FitzPatrick paid fleeting attention to all the "Bring Your Stories to Life" Ancestry.com commercials and 23andMe.com's "Health Happens Now" jingle when they'd flash across her TV screen.*

*"Someday I'll get to that," thought FitzPatrick, fifty-seven, a Lewes, Delaware, freelance writer and mother of four.*

*"I'd kind of dipped my toes in exploring around, and I did the Ancestry.com DNA test, but that was in the early days, before it exploded and everyone was testing their DNA. I didn't really find much and kind of forgot about it."*

*The term NPE (not parent expected), describing people who discovered for themselves that their father isn't their biological father, was a foreign term to her. Why would she know about* that? She and her three siblings were raised in a no-drama suburban Pennsylvania household by their loving mom and dad.

But sometimes it is other people who make the NPE discovery and who must then decide if, when, and how to share that news.

That's what happened when FitzPatrick received an email from Jeff, who had recently discovered he had a half brother he never knew about. Call it an NBE (not the brother expected) situation. His father, it turns out, had a couple of previous families and three children that he'd fathered, but he had not been involved in their lives and hadn't told anyone about his previous families.

The brother was Joseph (Joe) Thomas Wingert, Bridget FitzPatrick's father.

"I was 'found' by an uncle I did not know even existed, who found my siblings and me through a popular ancestry database," says FitzPatrick. "So, although I was not the person actively seeking to mend any holes in my family's quilted history, I watched with joy as his questions were answered and other mysteries began to unravel."

The roots of the mystery reach back to her father's childhood in the 1930s, when his father, Jacob Wingert, suddenly disappeared from his life—and the lives of his mother and sister—just days after he was born.

"For the remainder of his life, Dad never pursued his own father's whereabouts," says FitzPatrick. "Perhaps it wasn't worth the heartache, or maybe he never felt the need."

There was no legal divorce or separation, just a vanishing act. In his absence, FitzPatrick's father was raised by his mother, maternal grandmother, and maternal uncles. The name Wingert was not quite abandoned, but he grew up nicknamed Joe Carney, his mother's maiden name and the surname of his Irish Catholic grandma and uncles.

Little thought was given to the German or Hungarian Wingert heritage. No one really knew, and it didn't matter.

Fast-forward to 2015. That's when the call came from Jeff, the half brother he never knew about.

"My father was terminally ill with only weeks to live by the time he was serendipitously 'found' by his half brother," says FitzPatrick. "Dad surprised us all by expressing a willingness to meet him."

He said, "Sure, let's have a party."

The timing was imperative, as her father was being moved from the hospital into hospice.

With about twenty relatives gathered at her parents' Philadelphia home, FitzPatrick and her husband, Ward, went to the airport to pick up her new uncle from Omaha, Nebraska.

"When I first looked at him, the resemblance was uncanny," she remembers. "It was like looking at my oldest brother."

An only child, Jeff had been raised in Iowa and was only ten when his father (FitzPatrick's long-gone grandfather) died from lung cancer. Through Jeff's research, he discovered that the elder Wingert had a first family, where he'd fathered Joe and his sister, and a second family. Then he had divorced, remarried, and had Jeff ten years into that marriage. Jeff is only three years older than FitzPatrick.

"If you could have seen the look on my father's face when Jeff came through the door," says FitzPatrick. "He was glowing. I can't speak for him, but I think he was very pleased to meet his legacy, his half brother. Jeff is a successful, nice guy, and they really hit it off. It was a very, very happy day."

Perhaps one of the greatest benefits of their meeting was Jeff's ability to demystify the image of their grandfather, whom they'd previously seen as a bad guy who just one day took off and left the family.

FitzPatrick says that Jeff told nice stories about his father, and that it sounded like he had a very happy, idyllic childhood. Joe Wingert never knew he had other siblings, other than his sister Lenore, also raised by the Carneys. It turns out that their father, Jacob Wingert, was also an interesting and successful man. Their father, Jeff explained, was a mechanical genius who fixed factory equipment and even trained pilots during World War II.

"I really believe this brought some closure for my father," FitzPatrick says. "To think he was terminally ill, and a half brother walked in, it's like divine intervention or something. They liked each other right away and even had some of the same mannerisms."

Her father died three weeks after the visit.

"If we had put Jeff off, or waited, they never would have met," she says.

"Jeff's search was prompted only by his own children's questions about his ancestry," she says. "He was flabbergasted to discover he had a half brother and sister! All of this was discovered after being told this for almost sixty years that he was an only child."

Like many of the out-of-the-blue sibling DNA discoveries, this one is chock full of coincidences. FitzPatrick's middle son, Matt, was attending law school at Creighton University in Omaha, just blocks from Jeff's office. When Matt graduated, just weeks after they first learned about Jeff, members from both families got together to celebrate and learn more about each other.

"We're staying in touch and connecting at holidays," says FitzPatrick. "Ours is a very happy family in the fact that my father and all of us welcomed the news. Well, at least eventually. At first, my mother was a bit protective of my father, especially given his health at the time. But [Jeff] turned out to be an interesting and kind guy and a wonderful new addition to our family."

Meantime, the reunion has sparked FitzPatrick's interest in ancestry research.

One mystery that has been resolved: the Wingert family is not Hungarian, as was assumed. In fact, the descendants are predominantly Irish, Scottish, and English, with heritage from a smaller fraction of multiple northwestern European countries.

She says, "My father was able to heal an open wound before he passed away, and my uncle has become a close friend, especially with me. He got the bug that ancestry-seekers often relate, filling in the gaps of knowledge and lore. And although I found it hard to miss the grandfather I never knew but always wondered about (since his name branded my identity until I married), finding my uncle has revealed many wondrous stories about his adventures and service to our country."

Today, she's discovering what she calls "more leaves on that side of my family tree." They continue to be filled in, as do other branches from her mother's side—a family that has also dealt with painful separation through emigration and unknown destinations. "In my case," she says, "I have experienced both a sense of closure and a profound respect for my ancestors' journeys to the United States."

### LEGACY LESSON

"Hearing the nice stories about my grandfather has prompted all of us to not make assumptions and to give more thought to why he left. Maybe the Carneys, my maternal grandmother's brothers and family, kicked him out. We'll never know. But my father was able to connect with a part of his family that brought him joy."

### Upending the Family Tree: DNA reveals life-changing answer to "Who am I?"

*In her novel* Tree of Lives, *award-winning author Elizabeth Garden explores, among other things, fate and how we carry ghosts from the past with us into the future.*

*The story follows Ruth, a white Anglo-Saxon Protestant Connecticut artist with a rich pedigree. Ruth finds herself drawn to Jewish people in her community, the only people she came across who treated her with kindness, besides one grandmother with whom she was very close.*

*During her tumultuous life, Ruth briefly befriends a famous man who advises her to "marry a Jew." Later, she finally meets a very nice Jewish doctor and even converts.*

*Then she takes a DNA test and discovers she*

> *is 14 percent Jewish. It turns out, her beloved grandmother was Jewish but was adopted and raised by another family.*
>
> *The book has gained widespread attention.* Tree of Lives *won a gold medal in women's literature from the Florida Authors and Publishers Association President's Book Awards.*
>
> *But the plot thickens when readers discover the work is autobiographical and describes Garden's own journey as she uncovered a life-changing secret in her own family.*

Growing up in southwestern Connecticut, Garden, sixty-five, never quite felt comfortable with all the emphasis on her family's ties to the Daughters of the American Revolution and all the hoity-toity machinations required of families of affluence.

In the 1960s, her town was one of the nation's most patrician and palatial communities, and far from an ethnic hodgepodge. The town was home to restrictive country clubs, places where, by many accounts, Jews were not always made to feel welcome, and Garden never quite accepted that she was solely a descendant of the Puritan settlers who ran the show. Instead, she was always drawn to her Jewish friends.

Then, a family secret emerged—her father had committed a murder after escaping from a mental institution.

It burst the bubble that surrounded the family's serene WASPy perfection.

It had always been clear that Garden was marching to a different beat, a trait that took her to a different place emotionally than the rest of her family and allowed her the freedom to follow her instincts.

"My house was so oppressive, but when I was with my Jewish friends, I came alive, I belonged," she says. At one point during her career as an art director, she crossed paths with the American director, writer, actor, comedian, producer, and composer Mel Brooks, and the duo clicked. Meeting over drinks, Brooks told her, "You need a Jew."

Shortly afterward she met her husband—her Jewish husband.

The sense that she was closely connected to a Jewish heritage lingered, and she decided to take a DNA test. She'd heard her lineage could be traced back to biblical days, so maybe she could even find a Jewish thread way back.

"The result was a lot more than a thread—it was a whole new warp and weft in the family tapestry," says Garden. She discovered she was 14 percent Jewish.

She learned that her beloved grandmother was Jewish by birth but had been adopted and raised by another family.

All along, Garden wanted a sense of belonging to a group that she wasn't part of but knew she was somehow connected to. The discovery brought her a sense of rooted-ness within a culture she'd always been drawn to without understanding why.

"I think inside all of us is a quest to look closely at what is right beneath the surface, to see who we are," says Garden. "And I think that if we listen closely enough, there are ghosts from our past—in my case my great-uncle, who was calling out because he wanted to be understood. Our ancestors need us to connect some dots, and only those of us who listen to their voices can do that. Otherwise we would be repeating their mistakes."

### LEGACY LESSON

Garden expresses the lessons she learned through her lead character Ruth in her book *Tree of Lives*. While Ruth successfully identified the Jewish connection she felt, she also learned more about the well-documented historic ancestors who preceded her.

"Ruth's legacy lesson becomes hardwired by the conclusion of her unusual life story, a future legacy written in the DNA of her grandchildren," says Garden. "Ruth saw her part in the making of this new type of world person as mission accomplished, her bit toward creating a world free of prejudice."

## Ancient Wisdom: A long-ago ancestor's journal changes the course of her life

*Who am I? Where do I come from? And where am I going? Those were the questions the always-inquisitive Caroline Guntur (then Caroline Nilsson) peppered her parents with from an early age on.*

*Born in Sweden, Guntur was an only child. She lost her maternal grandparents during her childhood. Her paternal grandparents weren't in her life, either.*

*"We had a seemingly very small family," she says.*

*Her mom's response to her constant barrage of queries to learn more about the family story was often answered with: "I don't know that much. My parents didn't really talk about these things. We were not the kind of family that was into all that feeling chatter. We are forward focused, not backward. What's past was history. No need to revisit it."*

*Years of persistence paid off when Guntur's high school teacher called on students to create a family tree and write about their family history.*

*"I was overjoyed," says Guntur. "My teacher thought it would help us to relate to history in a deeper way."*

*For the teenager, it was all the ammunition she needed to start cracking her family story open.*

19

* * *

Shrugging in resignation, Guntur's mother started pulling out memorabilia she'd kept from her own father's stash of memories in the family's living room cabinet.

A worn ticket stub for a steamship passage to America became the major clue to the elusive mystery.

"I got very confused, because as far as I knew, my whole family was Swedish, so why would they have a ticket like that?" she remembers asking.

Guntur would discover that her great-grandparents, Per and Anna Lasson, had emigrated to Duluth, Minnesota, in the late 1800s. Per had helped build the railroad in Duluth, and they stayed there for almost ten years before returning home to Sweden, where Guntur's grandfather (her mother's father) was born.

"This blew my mind, because I had never heard this story before, and it ignited my passion for family history," she says. "I started researching and never looked back."

As she kept researching, she found a journal from a man named Michael Tengwall, her six-times great-grandfather. In it, he wrote about his experiences in 1738 being shipwrecked off the coast of Cadiz, an ancient port city in Spain. The entire ship capsized, and Tengwall was one of the few survivors. (See following excerpt from Guntur's essay "Setting Sail: How One Man's Story Changed My Life.")

That journal completely changed her life.

"I found story after story about this amazing adventurer, my relative, and it changed my whole attitude toward life," she says. "I went from being insecure to understanding that I was really standing on the shoulders of some truly amazing people. Newfound confidence and a sense of adventure propelled me forward as never before, and because of this, I dared to emigrate to the US by myself at age eighteen with a few dollars and a backpack."

In 1999, inspired by her trailblazing relative, Guntur remembers literally spinning a globe to decide where life would take her next. It landed on Hawaii.

Soon she was off to enroll in Hawai'i Pacific University for her undergraduate degree, where she met her husband-to-be. The duo moved to Chicago and got married, and Guntur attended Columbia College for a master's degree in media management.

Today, she lives with her husband and daughter in Algonquin, Illinois, a far northwestern suburb of Chicago, where she is a professional genealogist and photo organizer who runs her own business, the Swedish Organizer LLC.

She attributes the unexpected discoveries in her own lineage to her passion for her current vocation and avocation.

In her research, she also found a female relative in the 1600s whose husband was a trader in the Baltic. But when he died, she jumped right in and took over the business.

"What better person to relay the emigrant tale to

descendants of Scandinavian migrants than someone who is one," she says. "I love helping people make their own discoveries. My family's story was a gift to me and had a profound effect. Seeing that they went all the way back to the Vikings and that there was so much courage and adventure in their lives helped me explore and own that part of myself."

## LEGACY LESSON

"This is such an exciting time for people seeking their ancestry, and I encourage everyone to give it a try. If we don't write our stories, they will get lost. You will discover something that will change your life in one way or another. I promise it will rock your world in a positive way. If more of us explored our histories and learned about the struggles people faced and overcame, I think we would have a more tolerant world."

### Setting Sail: How One Man's Story Changed My Life—excerpt from blog by Caroline Guntur

Great storytelling is a gift. Not only to you, but to those who follow in your footsteps.

I know this to be true because my sixth great-grandfather Michael Tengwall (1705–1777) left behind him a journal, written at the age of sixty. In it, he tells the story of his life in incredible detail. Not only does it include his day-to-day life in small-town Sweden, but it also has an amazing recount of his travels.

In 1738, he was appointed ship minister and destined to sail for Constantinople on the naval ship *Sweden*.

After leaving his family behind to go above and beyond the call of duty, his journey took a dramatic turn when the ship foundered and he was shipwrecked off the coast of Cadiz, Spain.

As I was reading about this horrifying experience, I was amazed at his resilience and courage. After losing many of his friends to the storm, he and a few others who survived were taken in by the locals, who graciously offered them housing, clothing, and food.

Fortunately, Michael was an educated man. He spoke Latin with ease and was able to communicate his way around. There was one man, Don Cesse, that Michael clearly bonded with despite their cultural differences. It was obvious that the mutual respect they had for each other translated well and formed a lasting friendship. Sure, there were moments of despair, as Michael doubted he would ever see his family again, but also moments of wonder as he explored the Spanish culture and spent time learning about their customs. After living in Spain for some time, his return to Sweden was arranged by the consul in Cadiz, and he eventually made it home to the family who had presumed him dead.

What an amazing adventure, right?

I was so happy to find a firsthand account of my ancestor's life. Michael knew how to write captivatingly, by describing the sights, the smells, the textures of different fabrics, and the sounds of the church bells in the large Spanish coastal town. It was all in the details, and they were what really drew me in as a reader.

What was even more impressive was the balancing act he pulled off when writing this journal, because not only did he describe his travels so well, but he also somehow managed to describe his everyday life back in Sweden with the same amount of enthusiasm.

His trip was clearly a defining moment in his life, but it wasn't the only defining moment. He poured his heart out in this journal, explaining how heartbroken he was when he lost his wife to disease, when his little son died in infancy, and when his daughter lost her life at the age of six. He was surprisingly honest about how he "somehow found the strength to continue living through all the tragedy that had befallen him." His words made me wish I could have met him. It was absolutely fascinating to see just how fast life can change, and how fast the world keeps changing for all of us.

Reading this journal changed my life.

When I read Michael's journal, I realized that I wasn't the only person (and certainly not the first) in my family to have a thirst for knowledge and adventure. My bigger dreams started to make sense, and it gave me a sense of belonging and confidence that I had never felt before. It made me fearless and ready to see what would happen if I took a leap of faith. As you may have guessed, this newfound confidence completely changed the course of my life. Instead of staying in Sweden, I moved to Hawaii, met my husband, got married, moved to Chicago, started a family, and started my business in the family history field. That's what good storytelling does: it takes you on a journey and then comes full circle when you learn something new about yourself. Great storytelling inspires, and it changes lives. If Michael hadn't written in his journal, I may not have been equipped with the same attitude toward life that I have today.

## One "Shaky Green Leaf": Ancestry.com clue helps "bring runaway slave home"

*During the day, Taneya Koonce makes her living researching medical breakthroughs, clinical trials, and important information for patients and physicians as the associate director of research for the Center for Knowledge Management at Vanderbilt University in Nashville, Tennessee. But almost every night when she returns home, she hits her computer and goes into private detective mode, building her family tree on Ancestry.com.*

*Researching her family's lineage has been her passion since 2005. She quickly transformed her commitment to her avocation in becoming a dedicated volunteer in myriad local and national genealogy organizations. She's president of Nashville's Afro-American Historical and Genealogical Society, a member of the Tennessee State Library and Archives, and assistant state coordinator for the national USGenWeb Project.*

*It took her almost fifteen years to find the missing clue that she didn't exactly know she was hunting for but that had been intuitively inspiring her relentless digging. It was one animated shaky green leaf (the icon for Ancestry.com hints that pops up to help seekers discover new information) that broke open the most intriguing and "overwhelming and impactful" clue ever, she says.*

Koonce, mom of a fifteen-year-old daughter, says her fascination with her family story started when she was in college. During the summer of 1995, she spent time sitting beside both her grandmothers and taking notes about their lives. Over the years, both of them were diagnosed with Alzheimer's disease, and the story capturing began to fade.

"I kept the notes, but there were many holes in their stories that I wanted to complete," says Koonce.

That's when she, in her own words, "became addicted"

to Ancestry.com, FamilySearch.org, 23andMe.com, and other ancestry-seeking tools.

She rolled up her sleeves, hunting for information about both grandmothers, who were from Plymouth, North Carolina. Koonce was keenly aware that her family's lineage would be a story of civil rights and enslaved peoples.

Her search traced back to the 1800s and uncovered an ancestor of her second great-grandmother, Martha Jane "Mattie" Walker McNair. The ancestor was Martha's grandfather, Prince Walker, born in 1839 in Plymouth, Washington County, North Carolina.

One of the pieces of information she uncovered was Prince Walker's 1899 death notice in the local newspaper. The finding was a startling one, she says, because during those times, it was not as common to find an obituary written about a Black man in a rural newspaper. The newspaper article, reflecting the blatant racism of the times, called this relative a "before the war darkie." As she read this language, she understood that even though Prince Walker was enslaved, "he was considered a socially acceptable enslaved person by the white people then. They trusted him."

For years, Prince Walker was where the trail stopped.

Until 2019, when a shaky Ancestry.com leaf popped up on her computer screen.

"I was on the website working on my family tree when the hint showed up that someone else had Prince Walker on their tree," she says. But on this tree, Prince Walker had a son, Prince Walker Jr. (also known as John Prince Walker). The family details matched, including dates, names, and locations; this other family tree even had the obituary of Prince Walker (the elder) attached to the tree. Koonce realized that "apparently, this was indeed the same family" as her very own.

Koonce contacted the owner of the newly found family tree and discovered they were fourth cousins. That's when the story began to unfold—stories the cousin had learned through oral history passed down by her ancestors.

She shares the story: John Prince Walker, the son of Prince Walker, was sold to the plantation next to where he had grown up in Plymouth, where his father still lived. He tried to escape multiple times. Eventually at fifteen he escaped and made his way to Rhode Island and freedom. There, he enlisted as a Buffalo soldier, an infantry unit made up of African American soldiers who mainly served on the western frontier following the American Civil War.

"What was so amazing about his story is that his father had harbored him in his cabin so he could make his escape," says Koonce. "At the same time, his father was considered one of the slaves the plantation owners trusted, but he became this hero for his son. Somehow he was able to navigate his relationship with his owners and at the same time help his son toward freedom."

The creativity and resourcefulness she's found in

Prince Walker and John Prince Walker's stories has had a dramatic impact on her life, says Koonce. She says she'd always been fascinated by her family history and very sad about knowing her ancestors were enslaved, wondering what it would be like to be owned by these overseers and live in squalor.

"I love that my hobby links to what I do professionally, and that learning my family's stories is teaching me so much more about what history was really like," says Koonce. "When you learn about your family's context in the larger story of history, it personalizes it in a very powerful way. I knew my ancestors must have been slaves, I'd read about what that must have been like in history books, but when I learned I had an ancestor who escaped all of that, it really put the background of what exists in so many African Americans' homes in context for me. I went from living in a family that just never talked about any of this, except to once in a while say our relatives were sharecroppers, to really feeling what it must have been like."

Through her own work uncovering her ancestor's remarkable journey to freedom, Koonce has been inspired to use her ancestry detective skills to help others research hardship stories. Currently she is involved in the Fort Negley Descendants Project—an initiative to document and tell the stories of those who sacrificed at Fort Negley, told from their descendants' point of view. Per the Vanderbilt University website, "The Fort Negley Descendants Project is an oral history digital archive aimed at preserving the voices and stories of the descendants of the African-American laborers and soldiers who built and defended Fort Negley."[10]

Her research is also having a ripple effect in her own family. Recently she received a call from her father, who had just attended a funeral for a cousin.

"My dad thanked me for helping him understand how important it is to connect with and stay connected to our relatives," Koonce says. Her brother also surprised her when he met with a relative, someone she discovered on Ancestry.com and connected with on Facebook. "He was on a business trip in Hawaii and saw we had a relative there, so reached out to meet him. It's fun to see how the connections just keep happening."

## LEGACY LESSON

"I think learning about family history brings people together and helps us better understand the political and cultural climate that inspired our own lives. I think it gives us a better understanding of what we think are our differences. Instead of dividing us, they can help unite us. We all can really appreciate what our families have gone through and take these lessons of perseverance and put them into play in our own lives."

## CHAPTER TWO

## SURPRISE! SURPRISE!

Unearthing startling familial
revelations as you discover
your roots

Stories of identity—who we are, where we've come from—are the most life-transforming stories of all. Unearthing our family narratives is a powerful way to bring them to life and show us how we are parts of a bigger story.

Now, during an age characterized by mobility and rootlessness, people are looking to their ancestors for answers more than ever. Ancestral and genealogical exploration is enjoying unprecedented popularity and sparking endless fascination.

These efforts are inspired largely by the growing

ranks of Hollywood celebrities who are trekking across the globe on their searches on the hit show *Who Do You Think You Are?* Prominent celebrities including Kelly Clarkson, Mandy Moore, Gwyneth Paltrow, and Sarah Jessica Parker are inspiring all of us to launch our own poignant searches to trace our family trees in search of unlocking past mysteries.

In March of 2020, when the TODAY show's Carson Daly announced the birth of his fourth child, he said in an interview for NBC's NECN her name, Goldie, was inspired after he took an Ancestry DNA test he got from his sister for Christmas, which confirmed the family's 98 percent Irish ancestry. He said the results prompted him to search for Irish baby names, and on St. Patrick's Day he ran across the name Goldie. "We had a short list of names that we liked, but nothing was sticking its neck out," he told TODAY. "And I thought about it, you know, this is our fourth kid. She seems like the pot of gold at the end of our family rainbow."[11]

The popularity of at-home DNA testing kits gives us ordinary folk the opportunity to make our own amazing discoveries.

The good news: genealogy can enrich your identity.

But more often than test takers anticipate, these big reveals can be startling. Weighty family secrets can open a hornet's nest of long-lost relatives, newly discovered siblings, and far-reaching familial linkages.

What happens when you uncover buried family secrets? The man you thought was your father is not your biological parent, siblings are not your siblings, and your ethnic makeup appears to have been more a cultural practice than a genetic reality. Experts have begun to study the impact of these long-held secrets on those who were innocently hoping to learn more about their ancestors or genetic medical issues.

"There's a growing epidemic of surprises in families worldwide as ancestry searches and DNA testing kits are spiraling in popularity," says Anita DeLongis, PhD. Through a study at the University of British Columbia's Centre for Health and Coping Studies, she and her team are exploring this phenomenon in the hopes of finding answers that will help everyone (including the individual making the discovery and the entire family involved) cope.

While stories of these shocking experiences seem to be popping up every day in conversations among colleagues, friends, people at the gym, coffee shops, and support groups, and just about anywhere people gather, she says we're just at the tipping point.

"The number of these surprise ancestry and DNA findings is going to explode," she says. "Unfortunately, not a lot is being done to help people cope with this information, which can really shake up their lives. People need to know DNA tests are not just a fun holiday gift, but they can open a potential Pandora's box of pain."

These are tricky situations, and ones the people profiled in this chapter often say were prompted by a similar inner stirring: "I always felt something might be a bit off."

In this chapter, family detectives share stories of surprises, reunions, and personal growth that are the direct result of their emotional journeys to find out who they are and where they come from.

## Mamma Mia: DNA test reveals clues to biological father

*Kristine (Kearney) Celorio has experienced a whirlwind of adventures during the last two years. Married to Alejandro Celorio, a Mexican diplomat to the United States, and the mother of two young children—Alex, seven, and Audrey, four—Celorio and her family moved to Mexico City from Washington, DC, in 2018.*

*That journey was chronicled on the popular HGTV show House Hunters International. Appearing on her favorite TV show was a dream come true for Celorio, who writes about the brush with fame in her blog, "Irish I Were Mexican."*

*"Being married to a diplomat, I knew one day I would move abroad," she says. "I told myself, 'When that day comes, I will apply.' So, a decade and two kids later, word came that our family was off to Mexico."*[12]

*Shortly after the move, she sent in a casting video, and months later, the family was selected. Five days of filming, and they made their TV debut in a half hour segment in October 2019.*

*Behind the scenes of the Celorio family's moment in the television limelight, her personal life also was playing out like a real-life episode of the celebrity genealogy drama* Who Do You Think You Are?

*Celorio had grown up with adoptive parents in a San Francisco suburb. After discovering and meeting her biological mother, she had exhausted all efforts to track down her biological father through online ancestry sites and 23andMe.com.*

*But the clues remained elusive.*

*On Friday, December 28, 2018, an emotional surprise came out of the blue. She remembers exactly where she was and what she was doing. She was sitting on her porch at her new home in Mexico when she received the text:*

*"I'm looking for my fabulous daughter, Kristine, you there?"*

*Celorio's response: "Huh?"*

*The next text: "It's your dad!"*

Raised in San Rafael, California, Kristine Celorio says she had "an amazing, idyllic childhood."

"I always knew I was adopted because my parents weren't able to have their own biological kids, but I never had a negative feeling about that," says Celorio. "My parents just told me my biological mom was young and adoption was a better option for me. I never felt abandoned, but I did always have a curiosity and a yearning to know where I came from."

At nineteen, during her freshman year at the University of California, Santa Barbara, Celorio did some sleuthing and began a traditional adoption search, only to discover hers was a closed adoption. She did learn she was born only miles away from the home she grew up in.

That was in 1989, long before DNA test kits were around, so she continued her search via traditional research. She uncovered her biological mother's maiden name and the fact that she had attended a private university nearby. After some digging for files and information at the university, Celorio uncovered her mother's address and discovered she was now living about twenty minutes away. She had a young daughter.

Celorio reached out to her biological mom. The two agreed to meet at a nearby restaurant. She remembers thinking the physical resemblance was undeniable.

"It was interesting to see she looks like me," says Celorio. "We have the same hair, but we have different eyes."

Her mom explained that she was raised in a conservative Catholic family and was a freshman in college when she got pregnant. In the eighties and nineties, sex, not to mention being pregnant, was not something she and other good Catholic girls could talk to their parents about.

"She told me she felt the best thing she could do to give me a full life was to give me up for adoption," says Celorio.

But the answer to the other question she had hoped to ask her biological mother—"Who is my father?"—remained a mystery.

Her mother did not know.

"It was great to meet her and kind of was just like meeting a stranger for the first time," says Celorio. "It's not like we embraced or were sobbing. We weren't overly emotional.

"Eventually I met her husband and daughter. It's not like we forged some great mother/daughter relationship, but every couple of years, we call, and it's just like I have this interesting person that I like keeping in my life. She told me she was unsure who the father was, and at that time I gave up all hope of ever filling that gap, of knowing where I came from."

"Two decades later, along came all these DNA tests, and I decided to take one just to find out about what region of the world my ancestry was from," says Celorio.

Fast-forward to Thanksgiving of 2018. Celorio had just moved to Mexico City with her family. She sent her saliva sample to 23andMe.com, hoping that it might turn up some clues. But she says she had no expectations. The tests revealed her mother's biological brothers.

In December of 2018, a DNA relative in Florida popped up who was related to her biological father on a cousin level. That set off a whirlwind of emailing. They discovered they all shared a family history in the San Francisco Bay Area.

Then one day Celorio was sitting in a rocking chair, having a cocktail with her family, when she got the text from her dad saying he was looking for his daughter. She remembers being stunned and texting him: "What is your name?"

Seconds later, the next text read, "John B—."

Celorio immediately did what every sleuth would do. She googled his name. The result—a white supremacist in jail for murder—popped into the search box.

"Yikes," she remembers thinking.

Moments later, a photo came from the John B— who was claiming to be her biological father. Thankfully, he clearly was not the murderer.

"Phew," she thought.

### *"The second I saw his picture, I knew it was him."*

"The second I saw his picture, I knew it was him," she says. The story he told her added up. He was her biological father. He was currently married with a child and living in New Zealand.

An hour later they were chatting on the phone, and

Celorio learned the details of her biological parents' relationship.

Her father had been living in Mendocino and doing woodworking. One of his friends introduced him to her mother, and the duo started dating. Every weekend he would make the two-hour trek to visit her at college. By the end of summer, the romance broke up. He never knew his ex-girlfriend was pregnant.

The names, the timing, and the story all fit.

When Celorio and her biological father exchanged pictures, the eyes told the true story. "I've got his eyes," says Celorio.

The two speak frequently, and John B— has told Celorio he would like her to meet his mother and larger family, who live on the East Coast. She discovered she is about 70 percent Irish and part Scottish and English.

## LEGACY LESSON

"Ultimately, finding my biological father has filled a small gap in my understanding of who I am," says Celorio. "I feel kind of relieved now that the picture is more complete. Suddenly, there are no more questions. For so many years, I had all those questions. Now, I just know a little bit more about me. But I'm still who I was before, someone who was raised in a very happy family. It's just that now I know more about this other piece that is part of my life."

## Anything Is Possible: "You and I are sisters!"

*Janine Dzyubanny was "in shock" when she got the message. She thought it must be a scam.*

*She and her husband were on vacation when she got a message from 23andMe.com: "You share 47.9 percent of your DNA with Jennifer Frantz and we predict she is your sister."*

*The Wi-Fi where the Johns Creek, Georgia, couple was staying was spotty, so she kept clicking the message repeatedly. "I showed my husband and thought it was crazy," she says. "He had done his DNA and found a bunch of fifth and sixth cousins, but nothing like this."*

*Then she sent off an email to Jennifer:*

*"Hi, my name is Janine and I just received my results that show you and I are sisters!"*

The message unlocked their history and connected their lives, unraveling an even more shocking series of serendipitous events. Both sisters and their families had been living thirty minutes away from each other in Georgia for over a decade. They had both given birth the same year and have fifteen-year-old daughters (Jennifer also has a fourteen-year-old son). They were both pitchers on their softball teams. Both families have the same breed of dog, and they were both raised as the youngest in their adopted families with older brothers. They both are preschool teachers. Both sisters have best friends named Katie. They both thought they had been abandoned on the streets of South Korea and taken to an orphanage for adoption.

And, for more than four decades, they had both been completely unaware that they shared half of their DNA with anyone other than the parents who put them up for adoption.

"I didn't grow up thinking 'poor me, I'm adopted,'" says Dzyubanny, who grew up with three older brothers. "I had very loving adoptive parents and always thought I must have a guardian angel who took care of me and made sure I was raised in such a great home. I just never had any interest in looking for my parents."

In 2017 Dzyubanny took a DNA test solely to explore her medical history. "I just wanted to find out if my diabetes was genetic," she says.

Decades ago, no one ever imagined that family secrets like this would be revealed by science. Neither did Dzyubanny.

When she was sixteen, her parents let her read her adoption file, which said that she had been taken by an orphanage in Seoul and adopted at five months by a family in New Jersey.

"I never thought it would have been possible to find my birth parents, so I didn't even imagine the possibility of a sibling."

In 2016, a year earlier, Frantz, who was born just a little more than a year before Dzyubanny and adopted by a family in New York, got a DNA test for Christmas from her husband, Ben.

"Being adopted, I had no idea of any health risks, and I kind of wanted to know for my two kids, too," says Frantz. She and her four older brothers and parents had moved to North Carolina, where she was raised, and she had moved to Canton, Georgia, as an adult.

Her DNA information revealed nothing startling, just a few very distant cousins. For months, she just got 23andMe.com notices of third cousins. "That was the closest to relations I got," she says.

That was until a year later, when Dzyubanny took her test, and the dramatic results sent shock waves through both their lives. Frantz was on vacation in Aruba and remembers being "shocked."

"Both of us are very different personalities," says Frantz. "Janine is the type who jumps right into the water, and I'm more cautious, the kind of person who dips a toe in. I remember opening the email from 23andMe. com while lying on the beach. I thought it was a hoax and must have sat there for two hours contemplating it. I went into the hotel, showered, and checked the 23andMe website and discovered it was real, and that's when I told my husband. He thought it was a hoax, too. Then I did some Facebook stalking and learned a little more about her, her husband, and her daughter, and that was kind of fun."

With one sister on vacation in Ireland and the other in Aruba, the duo began texting each other and made plans to meet.

In the summer of 2018, the two families met, and they have closely bonded. Since then they have met for breakfast almost every Saturday, they text and phone each other throughout the week, and they even launched a party planning business, Lucky Penny Planners, as their "side hustle." "It's just another way for us to hang out together and bond," says Frantz.

"I have the sister I always wanted, because I had three brothers," says Dzyubanny. "Of course, I never, never thought that was possible, but you know how kids imagine things."

The sisters have also delved deeper into their history. Through the agency that handled the sisters' adoptions, they learned more about their birth parents. They were dropped at the orphanage on August 23, 1976. Janine, named Mi Sun Han, was about five months old at the time. Jennifer, named Mi Kyung Han, was eighteen months old. Their parents were unmarried factory workers without the means to care for two children.

When Dzyubanny was adopted at five months, she learned, the agency accidentally swapped her file with the file of another baby who died during the adoption process.

All her life, she had celebrated her birthday on April 10. But after she discovered the agency switch, she learned her real birth date is March 5.

"It wasn't exactly an identity crisis, but it was pretty shocking," she says. "I never thought the baby picture my parents had of me looked like me. And now I know why. It wasn't me."

For Frantz, the discovery of her sister is one that was meant to be.

"It's just been so crazy, but I really believe that we were always meant to find each other," says Frantz. "Some higher force in the universe worked to pull us together."

Their sisterly resemblance is remarkable, both friends and family say. But Frantz, who is five-one, is quick to point out about her much taller sister, "I only wish she had given me some of her height."

## LEGACY LESSON

"Finding my sister filled a hole I didn't even know existed. I always felt lucky in my life and never even dreamed this could happen, that I could have a family out there. This just shows that anything is possible," says Dzyubanny.

## The Search for Self: An unforgettable road trip and a DNA test reveal startling results

*With more people using direct-to-consumer genetic tests to find relatives and learn more about their heritage, family secrets aren't staying secrets anymore.*

*Shortly after a significant vacation, Megan decided, "just for fun," to take a DNA test. She didn't expect any surprises. Her family's Irishness was completely entrenched in their daily living and identity. In fact, she and her husband had just returned from a milestone trip—a trek to her Irish ancestors' homeland.*

*A few months before taking the test, she had packed the Nikon her parents bought her for a birthday nearly twenty years earlier and headed off on a once-in-a-lifetime vacation to the Emerald Isle. From Dublin, to Galway, to Bunratty, Dingle, Kenmare, and Kinsale, the duo ventured along the Wild Atlantic Way southern countryside, soaking in the music, seafood chowder, scenery, culture, and Celtic people.*

*She chronicled the adventure in a gallery of photos on her blog, where she creates photo travel guides, saying that Ireland was even more magical than she expected, and she couldn't stop taking photos of the beauty of the Emerald Isle.*

*When she returned home, she sent her saliva in and, a few weeks later, opened the results.*

*Irish wasn't listed in there as a primary heritage. At first, she thought it must be a mistake, but then she dug further. Hers is a story of the risks and benefits of genetic testing.*

Growing up, Megan's family embodied their Irishness. Leprechauns, rainbows that led to pots of gold, and four-leaf clovers that brought luck connected them to thousands of people across the globe who are descended from the same proud clan stretching back for centuries.

Some of her memories included watching the Chicago River being dyed green and attending Chicago's annual St. Patrick's Day parade, one of the largest in the country, which connects families like hers to the patron saint of Ireland and their Irish heritage.

But beyond the commercialization of the family's heritage, she longed to immerse herself in the culture, music, and history of Ireland, and to follow the footsteps of her ancestors through the Emerald Isle. In May of 2017, she and her husband spent their nine days in Ireland. Her photos (handstands at the Cliffs of Moher) speak volumes about her adventure.

When she returned home, the DNA results changed her past.

She went on the 23andMe site to view her DNA

matches. "Odd," she remembers thinking, "there's no indication of Irish heritage." Even more confusing, her major heritages consisted of Danish and Swedish.

"At first I didn't believe it," she says.

In one second, her familial and cultural identity had changed.

"Later, I found out my dad (whom we grew up with, so naturally we thought we were Irish) is not my biological father," says Megan.

Megan, like a small but growing segment of young adults, had inadvertently discovered she was conceived with donor sperm. "My parents conceived me via artificial insemination," she says. Her 23andMe DNA results were the first clue that led her to discover her parents had undergone artificial insemination in a Chicago lab to conceive their daughter.

Though she says the discovery was startling, she knows that her parents were infertile and "did everything they could to have a baby," she says. But she plans to stop her DNA research there and has no intention of searching for biological siblings who have also done the genetic tests.

"It's certainly been hard at times, but it also actually solidified that family is about much more than blood," says Megan.

"At first it was quite a shock—everything became so technical. 'My dad's technically not my dad.' But slowly, after the shock began to fade away, I started realizing it didn't really matter anymore. I was lucky and fortunate to be raised in a loving, caring environment, with two parents who wanted the very best for me. As we continued meeting up for holidays and events and having semiweekly phone calls and doing all the normal stuff families do, it's kind of like that technical piece faded away and was enveloped by normalcy and love and support."

## LEGACY LESSON

"I think it centers around the fact that some people get technical about the data and really focus on chasing data and metrics. This is especially important when it relates to family. I think a good way to compare it is to liken it to people who are constantly looking at what they weigh on the scale instead of focusing on how they feel and look. Technically, I have another biological father, but in my view, I have a very special father who raised me."

### Family Matters: Study explores how to help everyone in the family when DNA tests uncover buried secrets

As a professor of psychology, Anita DeLongis, PhD, has devoted her life to studying the impact of stress on people's health and mood.

In the past couple years, the Vancouver director for health and coping studies at the University of British Columbia couldn't help but notice how the explosion in popularity of DNA tests is opening mega-stress-producing revelations, sometimes with detrimental family secrets that were never known before.

Though genealogy seekers most times don't go looking for scandal—they stumble across it by chance—the missile of information can ricochet through a family, causing massive devastation and heartbreak.

Consider what it feels like to discover you are the illegitimate child of your mother's affair, or that the man you called "Dad" all your life is not your biological father. Or that your grandfather secretly was husbanding multiple families. These and other family secrets are cutting across religions, social statuses, and races like wildfire, or so it seems to DeLongis. She knows that these discoveries are likely to cause wrenching dilemmas for the entire shell-shocked family.

Over the last couple of years, DeLongis and her colleagues have been hearing these shocking DNA stories echoing through their social networks. They have also been contacted by patients and others who were often in a fog, approaching them with their stories.

"I kept thinking, 'How many people are having this experience?'" she says.

So, she took her first informal poll. While seated with her colleagues for dinner at a restaurant following a conference in Montreal, DeLongis decided to ask her fellow psychologists how many of them knew someone who'd had a surprise experience.

All hands went up around the table.

A new study was born.

Her team decided it was time to study how this at-home DNA testing phenomenon was impacting one's sense of self and the family. They wondered, "What are the range of coping responses to this potentially stressful situation? How do families integrate the new information into the family story?"

Through the UBC Genetic Connection study, her team at the University of British Columbia's Centre for Health and Coping Studies, one of the world's top research universities, is examining what people expect to learn, including their reasons for purchasing a test kit. They are also seeking to understand the impact the test results have on both the person who submitted the DNA, and the entire family.

"This is all happening so fast and so new, we really don't have a road map for how to navigate these surprise revelations," says DeLongis. "People have different coping styles, and we need to recognize what they need when confronted with this blindsiding information. How do you cope when you just learned your dad is not your dad? Some people are monitors who want to know everything and dig deeper, but others are 'blunters,' who just don't want to know."

"There are studies out there that look at the impact on the individual, but we want to explore how these DNA surprises impact the entire family, and others as well," says DeLongis. "We want to look at what motivates a person to take the test, and what the impact is after they do."

Fueled by the easy accessibility and popularity of these at-home genetic kits, DeLongis says her team expects the study to have an impact on a significant proportion of people. "Our objective is to understand the goals people have in seeking out genetic testing and the impact of receiving the results," says DeLongis.

(In Chapter 10, read DeLongis's preliminary tips for preparing for and coping with DNA discoveries. She says these will evolve with the findings, which are expected to be complete in summer 2020.)

---

### The Big Thrill: A behind-the-scenes glimpse into the evolution of DNA testing

We are living in an extraordinary time when anyone curious about their family ancestry can discover family records on the internet and take a DNA test to find the mapping of our human genome, which links us to our ancestors' genetic past—all without leaving home.

"We're at a tipping point," says Diahan Southard, founder of Your DNA Guide, a southwest Florida full-service genetic genealogy education firm whose experts help turn DNA test results into meaningful family connections. She also is a featured writer for *Family Tree Magazine*, the author of several self-published quick guides, a regular contributor for *Genealogy Today*, and the featured speaker for RootsTech Salt Lake City 2019 and 2020.

The field of genetic genealogy is relatively young but evolving quickly.

"Not too long ago, people would find a list of third cousins or even someone more distant, making it very difficult to figure out how you are related—there are just too many possibilities," she says. "But now, there's an explosion of DNA test takers, so there is much more information readily available, and finding a connection is almost inevitable for many people."

One question Southard is frequently asked is "Why should I take a DNA test?"

Her answer: "DNA gives you a record of your family, just as birth certificates or death certificates tell you information about your ancestors. Unlike other certificates, DNA documents your genetic history."

### How DNA testing for family history works:

- People purchase one of the genetic testing kits and have it delivered to their home. Major companies that offer these DNA tests include Ancestry.com, 23andMe, FamilyTreeDNA, and MyHeritage DNA.

- The only thing you have to do is take a cheek swab and mail it in with the return kit that is included.

"The results will tell you two major things," says Southard. "People who share your genetics and the places you and they are from."

### Southard explains more about what the testing tells us:

- Information about biological ancestors in your genetic tree about four to five generations deep (this is only available if they too have taken a test). This can help you fill blank spots in your family tree.

- Where your family came from in the past few hundred years in an ethnicity pie chart that reveals the percentages of each ethnic demographic.

While genetic genealogy combines the traditional disciplines of genealogy with these DNA tests and tells us much about the genetics individuals share, genetic genealogy also is gaining popularity with law enforcement personnel, who are using the test results to solve crimes, says Southard. The field is known as investigative genetic genealogy, or forensic genetic genealogy, she says. How it works is that DNA found on victims can be traced through the massive global library of genetics now available from these tests.

One of the highest-profile cases happened in April 2018, when California detectives used DNA to discover the identity of a man they allege to be the East Area Rapist/Golden State Killer in Sacramento. A serial rapist and murderer, the Golden State Killer had escaped capture for decades. The police plugged the DNA evidence from one of the crime scenes into a public DNA database and found several distant matches, which they used to build a family tree for the unknown suspect. When they narrowed the case down to the suspect, they took a piece of tissue from the trash outside his home,[13] compared the DNA, and found it matched with the DNA from the crime scenes, according to an article in NPR. They arrested the suspect at his home.[14]

A Colorado teen's murder, which had remained a cold case for thirty-eight years, is another high-profile case that police authorities cracked through DNA evidence. Her murder also was solved by comparing DNA from the crime scene to a genealogy database, after which it took only three months to solve the crime.[15]

Beyond our searches for maternal and paternal lineage, genetic genealogy is gaining popularity to help people make informed health decisions based on their genes, as we'll discuss further in Chapter 5. And companies offer genetic counseling for those who discover they carry risky genes. For example, American actress Angelina Jolie, whose mother died of ovarian cancer, opted for a double mastectomy after discovering the faulty gene BRCA1, which increased her risk for breast and ovarian cancer.

But DNA genealogy probing is currently mostly "an American phenomenon," Southard says.

"While the number of people being tested in other countries is growing, right now it's a very American thing (because of all our immigration)," she says. "People in other countries know where they are from, or they think they do, anyway, so they just don't see the value in the testing."

The future of this trend globally will have a lot to do with observations made by other cultures as to how this plays out in America.

"Europe and the rest of the world is keeping a close eye on how Americans are using DNA testing for law enforcement and that 'wow, I found my family' factor," she says. "For us, it is so tantalizing to pay $100 and get all these clues to our ethnicity that the novelty and possibility of it all outweighs other possible concerns that may give those in other places pause."

### Sister Act: Raised to believe her grandparents were her parents, then she learned the truth about her real mom

*All of her life Mary Beth was raised to believe her grandparents, Frank and Elizabeth, were her parents.*

*But it wasn't until she was in her fifties that she learned the truth. When she obtained her birth certificate for an international cruise, she discovered that everything she thought she knew about her family was a lie. Her parents were her grandparents, and her big sister, Elizabeth, was her biological mother.*

*Mary Beth spent her entire childhood believing that the reason she instinctively felt so different from her two older "spinster" sisters and brother was because of their significant age difference.*

*She was stunned. Her whole family had lied to her throughout her entire life.*

*If her sister was her mother, she wondered, who was her father?*

*Her mother, Elizabeth (who she thought was her sister), went to her deathbed not revealing who the father was. She would only say, "We could not get married."*

*Along with inklings that something was just*

*not right, her life had been peppered with clues she would piece together after the revelation. The name Elizabeth had been passed down for three generations—Elizabeth, grandma; Elizabeth, her sister (now discovered to be her real mother); and Mary Beth, short for Elizabeth, her name.*

*For Mary Beth, there were more revelations to come.*

*Like many families, the secret keepers carried the burden for decades. The impact of this sort of secret can have fallout for generations to come. Susan, Mary Beth's daughter-in-law, is speaking out to help other families know they are not alone, and because, she says, "the surprises keep coming."*

Fast-forward thirty years, and Susan and her husband, Mary Beth's daughter-in-law and son, thought they had a rough clue about what they would find when they sent away their DNA test.

"We assumed that we would find out the father was a married man," says Susan, an author in Atlanta.

When they got the results, they were astonished, she says.

Turns out her mother-in-law's DNA came back 50 percent Ashkenazi Jewish, and her husband's 25 percent. They'd been raised thinking they were both Irish.

"So now we know that the man that the Irish Catholic girl could not marry was Jewish," says Susan. "We still do not know who he was, but we've narrowed it down to five men."

Ironically, their father-in-law, Mary Beth's husband, had his own parenting puzzle. He had been raised by his aunt and uncle and did not know his true parentage, either.

The "whys" about the reasons Elizabeth and her parents kept the secret from Mary Beth spilled out in a complex narrative about her sister's colorful career serving overseas during World War II. She also believes the parentage was invented to protect them all from the societal mores at the time, which would have shamed Mary Beth for being illegitimate.

Her sister/mother had a fascinating story.

Elizabeth graduated from nursing school in New York City in the early 1930s. She served as a nurse for ten years before she volunteered for the Army Nurse Corps in 1942.

Elizabeth had fond wartime memories of Generals Bradley, Eisenhower, and Patton. She knew all three men and even had dinner with them. "He had this reputation of being very rough. He was actually a very understanding man," she said of General Patton. "He was it. A fine, fine general," she said of General Omar Bradley. "He was tops. Real good," she said of General Eisenhower. "We [the nurses] were always treated as ladies."

When the Korean War broke out in 1950, she served in Korea and was transferred to Yokohama, Japan, where she helped organize an evacuation hospital for wounded soldiers. For two years, Elizabeth oversaw a six-hundred- to seven-hundred-bed hospital in Yokohama. There, the wounded from the Korean battlefields were flown in and treated, only eighteen hours removed from the scene of the conflict. She said the men came in at a rate of 250 a day. The wounded were rapidly evacuated from Japan and returned to the United States.

In the late 1950s, she served as chief nurse at a hospital in Frankfurt, Germany. She served as a chief nurse at an army hospital in New York until her retirement in the early 1960s. Her rank at retirement was lieutenant colonel. As an army nurse, she performed her duty on three continents and served overseas during two wars.

## LEGACY LESSON

- **Silence sometimes is golden.** According to Susan, "We learned that a whole community could work together to protect a child. That is how we look at the lie that was told to my mother-in-law. She was protected by her grandparents, her aunts and uncles, her cousins, the parish priest, and others who knew. Many people must have known that she was illegitimate. But they never told her; they let her be raised in a Catholic home and attend Catholic schools, preventing her from being stigmatized as she most certainly would have been if her situation was known."

- **Don't judge a book by its cover.** "We all assumed my mother-in-law, my husband, and his siblings were Irish Catholics. Well, they are Catholic for sure, but their heritage is significantly Jewish. We are grateful to know their heritage."

- **The truth is not told.** "We also learned that not everyone is as interested in DNA as we are. We have family members who we know of due to DNA testing, who have shown no interest in meeting us. We must respect their decision and have not reached out to them aggressively. We'd love to meet them but will wait for them to make the overture."

## CHAPTER THREE

## HEALING:

### Finding Grace in Genealogy. Looking to the past for understanding

For many of us who search for our ancestors, we find ourselves discovering events from our heritage that keep repeating themselves throughout our family members' lives and across generations. Many of these events have become the foundation for our lives in the present.

The expression "It's just history repeating itself" isn't far off the mark. These ghosts of hurt, betrayal, lies, and hardships—which defined the lives of our ancestors and the pain that has been carried through the generations—are often embodied in the energy that continues to haunt us today.

Whether it's addiction, anger, violence, or absence,

these destructive behaviors and personal battles can be passed on to you.

Many of the people interviewed for this chapter say they feel it was important for them to confront and try to heal some of the wounds passed down by their ancestors. They felt they had an important role in going through that pain, and not around it, to try to find some healing for the generations who would follow.

Yes, so many family secrets are rooted in shame about issues that define our common humanity, such as infidelity, hidden sexuality, abuse, racial or religious origins, or infertility. But experts say the best antidote is to tell our stories. By doing so, we can heal the wounds for our entire lineage—wounds that have been holding those who came before us captive for years.

The more we look back, the more we find. These lessons can be bittersweet, as out of the pain, we discover our ancestors' stories of grit, resilience, and the triumph of the human spirit to survive and thrive in the face of adversity. These stories can enrich our lives going forward and give us the strength to know that we, too, carry on our own long and winding roads.

The stories in this chapter show us that there is grace to be found in our burdened pasts, and healing in our futures. When we heal and transform the wounds we carry from those who came before, we are changing the trajectory of those who will come after.

## Secrets, Lies, and Twisted Tales: Genealogy bombshell leads to uncertainty about the past and seeking healing for the future

*When Rachel read* Inheritance: A Memoir of Genealogy, Paternity, and Love *by Dani Shapiro, she was inspired to explore her own family history. She made a beeline to her computer to register with Ancestry.com.*

*"I wasn't expecting to discover anything particularly startling," says the fifty-four-year-old East Lyme, Connecticut, resident, yoga instructor, and owner of Yoga Keeps Me Fit. "My interest was in learning more about the people's stories and how they might relate to mine."*

*Her research dropped a bombshell that changed everything. She learned that her grandfather had many other secret families.*

*But the secrets didn't stop there.*

*Rachel discovered that her mother had carried and endured the damage of those secrets throughout her life. Almost as if a lightbulb was turned on, Rachel got a glimpse for the first time into her mother's erratic behavior.*

*That trauma had rippled through her grandmother's, her mother's, and now her own life in subtle and not-so-subtle ways, surfacing in a legacy of loss, lack of affection, hostility, and mistrust.*

*Rachel realized it was up to her to prevent future emotional damage to her family. To do so, she started unraveling the secrets, one by one. The extent of the delusion and deceit within her grandparents' relationship was shocking.*

*Rachel and her family's story is one of many examples of how the effects of trauma do not always dissipate, but rather can trickle down from generation to generation.*

Years before she launched her Ancestry.com journey, Rachel, who grew up in Kent, on the southeast coast of England, was handed a list of names and dates from her family lineage going back to the 1600s by her mother.

She also recalled how her mother had always told her that she was "sure her father was not her father," because he had treated her so badly. Rachel's mother told her she had repeatedly confronted her own mother with her questions about her father not truly being her biological parent. Her queries were met with radio silence or, "Of course he's your father."

"As you can imagine, this did nothing for their relationship, and [my grandmother] could not admit to having been with another man or her reputation would have suffered," says Rachel.

Through her research, Rachel discovered her mother's inklings were correct, and that her biological grandfather had abandoned her grandmother when she was pregnant (or shortly after her mother was born) and then went on to marry a woman in another town. Later, they discovered he had four other marriages, and he was arrested for bigamy and went to jail. Her grandmother then remarried a man "she didn't truly love and who resented my mother for having to do so because she was pregnant with her."

As a result of her traumatic childhood, Rachel says her mother struggled all her life with low self-esteem and depression, "which led to us having an extremely difficult mother/daughter relationship."

"We have undergone years of not speaking," she says. "During her worst depressive years, she wanted nothing to do with me and disowned me. She unwittingly perpetuated the very situation she endured herself as a child."

"My grandmother lived in fear of having the truth come out," says Rachel, which she says is highly understandable considering the cultural climate in the 1940s.

While it took years, and many unsuccessful efforts, for Rachel's mother to prove her theory about her father not being her biological parent, Rachel's own internet research was almost instantaneous: the biological grandfather was a bigamist.

### Shocked by the extent of the delusion and deceit
"He had been married to five different women," says Rachel. "His marriage certificates were under variations

of the same name. I also discovered amongst my mum's files a newspaper clipping from the *News of the World* about his trial and time in jail for bigamy."

The trail of lies kept going.

"My grandmother hid the truth, saying her first husband [the biological grandfather who ran away] was away at war [instead of telling people he ran away and turned out to be a bigamist], although he was residing in jail," says Rachel. "Recently I discovered a letter from her to my mother warning her not to dig into the past, her husband had done some 'silly' things that should be forgotten, she had worked hard to put the past behind her and would not help bring it to light."

"After watching the PBS series *Mrs. Wilson*, where the poor widow discovers her husband had been married several times and had children with each wife, I got a glimpse of how painful and shaming it must have been for my grandmother, especially in the 1940s, when women were totally dependent on their husbands for survival and a good name," says Rachel.

"My mother believes [my grandmother] found herself pregnant with nowhere to turn except to marry to protect herself and her unborn child, although there was confusion as to who the father was. I feel so much compassion for my grandmother at that time."

### *Healing the family trauma*

"Knowing my mother's past, I can see she put so much of her hatred and anger [toward] her mother onto me," says Rachel.

A mother of two young adults in their twenties, Rachel knew it was up to her to break the generational pattern of hurt, and she needed to try to mend her relationships. "My mother had no role model or confidant and knew no other way to be," she says.

As a result of these discoveries, she called her mother on Mother's Day 2019.

"I was anxious about calling, and I was unsure how I would be received or even if she would speak to me," she says. "I wished her a happy Mother's Day, explaining my research of our family tree using the information she had given me years ago, and I shared my discoveries of numerous marriage certificates."

The overture began the healing.

"She was extremely happy I had reached out," says Rachel. "Her own mother and sisters had always denied the facts, which I would never have found without her detective work years previously. She knew I believed her. It's sad that such a division existed between the three sisters on this subject—the full truth had been hidden from them as young women.

"I could hear it in her voice, the relief of validation. I know I will never have the relationship that I longed for with my mother. There is, however, a new understanding and forgiveness. I have an acceptance of the past, the sins that were committed, and the love withheld from generations of children that had no voice or protection. I know we both felt a new connection, as if something between us had shifted for the good," she says.

Despite the painful discoveries, she is grateful she pursued the search.

"I am so glad I had the courage to broach the subject with my mother, and [to have] shared our detective work decades apart," says Rachel. "She knows I know the truth. My mother's story is important; all our stories are important. Speaking our truth matters. The truth can and has brought healing to our relationship."

In that phone call, a rift of a lifetime had been breached, a spark of healing begun, she says.

She also reached out to her cousins, who are her mother's sisters' daughters, and shared the truth with them. "They were aware of some of the past—what their mother had dared to tell them. [Our] grandmother had forbidden any mention of it, and once the bigamy story appeared in the newspaper, she never allowed a Sunday paper in the house again," says Rachel.

"Through one man's actions, so many were hurt, but through finding and sharing the truth together, we have opened the way to heal," says Rachel. "My mother and I are now in regular contact, and I have had several letters from her telling me how delighted she is that we are so much closer, that in revealing the truth I have lifted a huge burden that she carried her whole life. This really is a miracle."

## LEGACY LESSON

"This has healed me—knowing that I have learned and understood the truth in our family," says Rachel. "I have learned how trauma can and has traveled through the generations, and that speaking the truth and owning our stories can heal the past and bring hope for the future."

She says it has also taught her how to communicate more effectively with her mother and others by setting stronger boundaries for her own emotional protection. And her ancestry search has brought "grace and nonjudgment" to her, she says.

"I can't be sure that I wouldn't have tried to hide the truth, either, if I had been in the 1940s with no way to support myself," she says.

Most importantly, she says that she has learned that kindness and grace are the way forward, especially with kindness to oneself.

"I choose to believe that we are all just doing the best we can in life," says Rachel.

## Making Peace With His Ancestry: Common heritage bridges Middle Eastern conflict

*When Hazem Diab and his family members simultaneously did their DNA tests, they were interested in learning more about their genetic makeup and discovering the whereabouts of relatives his parents had left behind when they moved to the United States from Palestine in 1969. Their inquiring minds also wanted to know where his sisters, brothers, and mom got their blue and green eyes.*

*The last thing the family expected to discover was that their Middle Eastern ancestors had Jewish ties. The light-colored eyes had their origins in Uzbekistan in Central Asia, which was also discovered in the family's DNA.*

*A jovial father of one who is a senior quality manager in Rockford, Illinois, Diab reports one of his family members said upon the revelation, "What do you expect? We've lived with Hebrews for many generations."*

*Diab shares that his mother, Samira, also expressed much compassion for the connection: "As Arabs and Jews, we all are descendants from Abraham. We have lived side by side, and therefore it is all good."*

*Now, thanks to the proliferation of DNA testing, it's more commonly acknowledged that genetic ancestry connects those involved in the Israeli-Palestinian conflict, one of the world's longest-running conflicts.*

Born in the United States in 1970 shortly after his parents moved to the Chicago area from Palestine to "pursue a better life," Diab is one of six children. He has studied in and returned to Palestine a couple of times, and he speaks Arabic. This summer he plans to visit with his twenty-year-old son, Tyseer, who is studying conservation biology in college.

Diab believes it's very important to keep the family heritage from the Middle East alive and is especially excited about showing his family the rich and historical archaeology of Israel, Palestine, Jordan, and Dubai. His father, Mohammad, retired in Palestine, and "I think this could be the last chance for my son to visit his grandfather there and really soak in the culture and his ancestry."

The DNA tests the family took that led to the "interesting" revelation was part of a "fun thing we thought we'd do to compare our results at a family reunion we were planning," he says.

He and his five siblings live in different parts of the country, including Illinois, Washington, and Colorado. They agreed to meet in Oregon "for the DNA reveal," and each took different DNA tests, including Ancestry.com, 23andMe.com, and *National Geographic*'s Geno test.

For the lone sister who opted not to embark on the testing, Diab and his siblings had a little surprise. "We told her to close her eyes and reach out her hand. We then put a package of Hebrew National hot dogs in her hands and said, 'Congrats, we're Jewish.'"

All in good fun, Diab says. The DNA testing has taught him some important lessons.

"Despite our differences and all the conflicts, we are all really the same," he says. "There was a *Newsweek* magazine cover a while back that was a split photo of a Palestinian woman and an Israeli woman, and they looked [almost] like twins. Through our tests, this came to life in my family, how we all are really alike."[16]

## LEGACY LESSON

"I learned that it doesn't matter where you come from—you control the quality of your character. You can't use the excuse that you have a hot temper because you came from Sicily, or any other stereotype about a nationality or part of the world your ancestors are from. But researching your ancestry can give you closure and help you better understand a little bit more about what the people who came before you had to go through. It's a funny thing, but I have a neighbor here [in the United States] who I really like a lot and am close to. When he did his DNA, he discovered he has a little part of him that is from the Middle East."

## The Parent Trap: Healing the sins of the fathers

*Jaimie grew up in a three-generational household with her parents, grandparents, and brother in a small town in Florida. Her family coped with myriad health problems and often chaotic behavior, but she describes herself as "the always-cheerful, easy child, whose needs weren't as important as everyone else's."*

*A distant, unattached relationship with both her parents has always been very painful for her. But when she started researching her family ancestry through FamilySearch.org, she started discovering clues to the family angst—census records that pointed to a long legacy of abandonment, lies, and emotional deficits.*

Jaimie, a blogger and international charity worker, lives in the Middle East with her husband. She developed an interest in ancestry when she and her husband started thinking about having children.

"I wanted to be able to provide a rich tapestry of history for my child, if I ever have one," she says. "So I started researching my own family tree."

The surprising documents she found during her search included records of her maternal grandmother living in an orphanage, even though her biological parents were alive; a marriage record of her paternal grandfather lying about his age and marrying a woman she'd never heard of before; and a complete absence of records for her paternal great-grandmother under the name she knew.

Armed with the evidence, she began quizzing her parents about these discrepancies.

No one had ever spoken of them before, but suddenly the stories started to pour forth.

Her paternal great-grandmother was a wild flapper girl who "couldn't be bothered to take responsibility for raising her son [Jaimie's paternal grandfather], so he grew up with his grandmother," she says. "She in turn was a spiritualist, who would communicate with the spirits, conduct strange rituals, and bandage her grandfather's cuts and scrapes with cobwebs."

Her grandfather left home early, lying about his age to join the military during WWII and again lying about his age at his first marriage—which she says ended so badly that she never heard about it, despite knowing him very well.

"I always assumed my grandfather was a distant, detached man because he was a war veteran—but this history made me realize his emotional deficit began long before the war," she says.

On her mother's side, her great-grandparents had struggled to raise eleven children during the Great Depression. Due to financial difficulties, they placed the youngest children (including her grandmother) in an orphanage. As their financial situation improved, they would take their children home, one by one. But her grandmother was left in the orphanage the longest.

"She carried this trauma with her for the rest of her life, becoming an unstable and detached mother herself," says Jaimie.

These findings were eye-opening, creating a space for empathy and healing.

"As an adult, I can understand my parents better when I realize the immense challenges of their own parents," says Jaimie. "My detached dad and detached mom both had parents who didn't know how to parent them. Understanding this legacy of pain and neglect gives me more mercy and compassion for their weaknesses."

Although she has discussed the findings with her mother and now has more empathy for the pain her mother endured growing up, she says, "Most of the impact has been on me personally. The Bible says that God visits the mistakes of the parents upon the children unto the third and fourth generation (Exodus 20:5-6), and nowhere is this clearer than intergenerational ancestry histories. I became interested in my ancestry because I want to have children but am paralyzed by fear that I will continue that chain of neglect and unattached parenting."

She explains that the Christian scriptures do not depict God as gleefully enforcing generational punishment, but rather that God respects human choices and allows the cycle of cause and effect to be played out in history. "Even though God allows natural cause and effect to happen," says Jaimie, "I love that there's always an 'escape button' with God. He tells us that 'The son shall not bear the guilt of the father, and the father shall not bear the guilt of the son.' If you live a life of faith, there's always a way to stop that cycle."

In many ways, the pain she experienced as a child led her on a spiritual path that has inspired the faith-filled life of volunteerism she practices with her husband in the Middle East.

"I was the kid who ended up alone with my thoughts," she says. "But in my aloneness, I found God was there."

As time went on, Jaimie developed a desire to follow a career that would prioritize her belief in a divine being who loved her and other vulnerable members of society.

She and her husband moved to Beirut, where, among other things, she taught Bible classes at a private school for Lebanese Christian students and worked extensively in relief programs for Syrian refugees, identifying some of the neediest families and delivering food packages on a monthly basis. Her work with refugees spurred her interest in studying the Arabic language and getting a master's degree in Islamic studies.

"It helped me enter their world," says Jaimie. "I spent a lot of time sitting on the floor of dingy basement apartments, listening to awful stories of displacement and vulnerability. It was in those moments of connection that I needed to be able to speak directly to the heart, without a translator, and give spiritual encouragement in terminology that would be familiar and meaningful to them." She currently is working on her PhD in religion.

## Distant connections

Beyond her immediate ancestry, Jaimie is putting together pieces of a more distant puzzle, which she says is a combination of "the good and the bad together."

For example, her nineteenth great-grandfather, Lord Randolph de Neville (b. 1295), was found guilty of incest with his daughter. On the other hand, her thirty-first great-grandfather was King Charles the Simple of France (b. 879), who was called "simple" because he was straightforward and honest and forged an important pact with the Vikings that ushered in a period of stability for the region.

Her eighth great-grandfather, Teague Jones (b. 1620), emigrated to New England at age twenty-five, married a Native American woman, and is remembered in court records as being an especially poor citizen, often in disputes with other settlers. At around the same time, her ninth great-grandfather, John Greene (b. 1597), was one of the twelve men who followed Roger Williams in campaigning for freedom of conscience in the colonies, later founding the city of Warwick, Rhode Island.

"Being able to step back and get a macro view of the

last generations helps me to step outside the picture and determine to do things differently," says Jaimie. "The third and fourth generation stops here. It stops with me. I think these findings have given me forgiveness for the weaknesses of my parents and more courage to consider parenting for myself."

She adds, "My family *experience* made me not want to have children, but after I researched and understood the *history* of how it got that way, that was when I felt encouraged to at least entertain the thought of starting my own family. Understanding how and why things got that way is what gave me the courage to move past the old mind-set."

### LEGACY LESSON

"Uncovering my ancestry has helped me to cultivate a mind-set of forgiveness and acceptance in a challenging family situation, because I realize that each family member was merely living out the legacy handed down to them," says Jaimie.

### Listening From the Soul: Tapping into your healing story

*All of our ancestors have a great story, a treasure for us to find, even when the events are traumatic and painful.*

*Bad things happen, but wonderful things happen as well, and we have a choice to listen, learn, and grow, says Judy Wilkins-Smith, a Texas-based systemic coach and constellation expert who works with clients to explore hidden patterns and unconscious personal issues in order to surface and heal unresolved trauma in their lives, often from family systems and ancestors. Genealogy sites show where someone belongs in the family tree, and systemic work and constellations help people understand how to use the family patterns to grow.*

*Generation after generation, our inherited emotional and behavioral patterns can keep showing up in our lives. Subconsciously we carry our ancestors' pain, language, actions, and even successes encoded in our energy fields, DNA, and cellular memories, which impact our health and our physical and emotional well-being.*

Through her work, Wilkins-Smith helps people see these patterns and behaviors and understand what they mean so they can mindfully change their language, actions, thoughts, and feelings and move in a new direction. To do so, she helps people rewire what she refers to as their emotional DNA and change their future. Systemic work and constellations are used by people in their personal lives, and executives in Fortune 500 companies also use them to explore complex issues and facilitate deep insights and breakthroughs, she says.

The way we carry what lives in our familial, cultural, and corporate systems can directly impact whether we succeed or fail, she says. Once we understand what lives there, we can also use this wisdom to create great success. Reframing "what is" into "what's possible" can often occur with insight and compassion once we understand what happened to those who came before us, or when we begin to move mindfully toward our own fulfillment. Failure to look at the past can result in repeating history. We may then live someone else's life as though it were our own.

"When we listen to the stories of our ancestors, as painful as they may be, we can create something that is beautiful," says Wilkins-Smith, who grew up in South Africa.

### Start listening to the language of the soul

"Ancestry.com is just the beginning," she says. "It tells you where you belong, but then you need to discover why that matters. Your patterns often begin in the mouths and bodies of your ancestors, until someone says, 'I'm changing.' When they know what they are changing, the shift is even more potent. Understanding how your ancestry matters puts a powerful tool into your hands, once you know how to use it."

When you make that decision, you become the primary pattern maker. And the more you know about your system and its patterns, the better you're equipped to create an incredible life and a different path—with acknowledgment of those who came before you.

"Change happens when you look into the face of family struggle and say, 'This doesn't work for me. I am changing the trajectory. I am doing this differently,'" she says. "These stories of significant events that happened to our ancestors create the mind-set that this is *the* truth. But it is not *the* truth, it is simply *our* truth, and you can change it."

Consider a woman Wilkins-Smith worked with who wouldn't get a driver's license because she was terrified she might hurt someone.

"She would get to the testing center and then go home," she says. "She did that five times. When I asked about her parents, it turned out her mother had killed a child who ran out in front of the car. She had never driven again, and now my client was carrying her fears as though they were her own.

"When she could see that the fears didn't belong to her and leave that with her mother, she could get her driver's license and along with it the freedom that her mother had given up. She also chose to teach road safety to kids at her local school to honor the child that had been killed," says Wilkins-Smith.

She helps clients see that they don't have to stay trapped in past lives of their relatives. "You can tap into the wisdom of your ancestry to start traveling a new journey. Often our greatest pain is our biggest gift."

Consider the loss of a significant member of your family, or perhaps a betrayal by such a person that spurs you into acting or moving in a new direction. If we look at these events objectively, we might say something like, "Because of this event or loss or betrayal, I did something different. Thank you."

**LEGACY LESSON**

"Pain can become a gift, and when we learn to look at our systems in depth, we understand how they are always in service of us. We are indeed living a remarkable life, if only we know how to see it."

---

### How to Break Multigenerational Patterns

**Exercise: Putting Patterns Into Practice**

Wilkins-Smith recommends this exercise to recognize patterns consciously, so that we can set down long-held struggles and create success, happiness, peace, and joy.

On a piece of paper, write down the one thing you would like to stop doing or experiencing in your life. Place it on the floor. Notice what you tell yourself and how you feel about it. You might even have a facial expression, a saying, or an action that happens for you as you consider what you wrote. Good. This is the pattern you are trying to stop.

Ask yourself: Are you the only one in your family who has had this experience or habit? Or is it just like your father, mother, or perhaps someone else in the family?

Now write down your heart's deepest desire or highest wish on another piece of paper. Place that on the floor next to the piece of paper that states the one thing you would like to change. What does that feel like when you compare the two? What do you tell yourself or feel about that? Where do you experience those emotions in your body?

Now compare the two pieces of paper. Notice how close to or distant they are from each other. Ask yourself: How much do I want this goal? Do I really want this, or do I just *want* to want this? Or am I even allowed to want this? This aspiration or wish is the pattern that is trying to surface in the system through you.

See if you can stand next to the piece of paper with your heart's deepest desire on it and notice how that feels, and where you feel it in your body. Can your heart open to that desire? If your heart can open, your mind will open, too. Can you take that step and be the change agent, breaking a cycle that no longer serves you and beginning a cycle of growth? *That* is the chapter that only you can write.

## Owning Your Story: Overcoming the shackles of ancestral shame

*The importance of being thin, pretty, and sexually attractive (but not too much) is just one of the beliefs Karen C. L. Anderson, fifty-seven, absorbed from her mother and grandmother while growing up in the 1970s.*

*For most of her adult life, she felt hurt and anger toward her mother because she internalized stories like "You're a pathetic loser" and "You're not worth anything," all because she wasn't "thin, pretty, and sexually attractive enough," according to her mother's standards.*

*She also discovered that those were messages embedded in the psyches of women throughout her lineage.*

*She knew she had two choices: stay imprisoned by them or transform those messages for her own liberation. She chose the second one, and it became her route to freedom.*

Today, she is dedicated to helping women gain autonomy and control over their own lives through revealing ancestral patterns, healing shame, and transforming legacies. In her writing, she challenges the images women have inherited from their own lineages about the way things are and can be for women.

In an excerpt from her blog, Anderson, author of several books on mother-daughter relationships and a master-certified life coach, offers these tips for embracing and telling your own stories—both the internal narratives you tell yourself and the stories about what you envision for the future and would like to see happen in your life.

"It's important to know what's your story and what's NOT your story," she says. "For example, things that happened to someone else are not your story to tell, especially not in detail. Make sure any details you share are pertinent to the narrative you tell *about yourself*."

From Anderson:

- **"Like and respect your reasons for telling your story.** This is an excellent gut check. The only person who must like and respect your reasons is...you. If there's any aspect of it that has you NOT liking or respecting yourself, pay attention to that."

- **"Stand on your story (rather than having your story stand on you).** This means you've transformed your story from being a source of suffering into a source of wisdom. You don't have to be quiet about what you experienced (in order to protect others), nor are you doing anything wrong by talking about it, especially if you can take responsibility—as an adult—for the way you feel about it. This isn't a resigned 'they were doing the best they could' statement. It's not about minimizing your experiences. It's about your resilience."

"I had a lot of negative stories about my own mother, growing up, and about myself," says Anderson. "But I have transformed those stories, so that I am no longer a victim to them."

## LEGACY LESSON

"There's a big difference between having once been victimized and living as a victim. And that difference will inform how you tell your story. How you relate to a previous experience is the difference between living in a 'less than' position and living with sovereignty. When you accept (which doesn't mean 'like' or 'approve of') your past and everything that happened to you, and everything you made it mean, without shame or fear, you help others do the same."

## CHAPTER FOUR

## CELEBRATING ANCESTRAL SURPRISES AND WISDOM:

### Embracing the gift of your ancestors' stories

To come to know and make sense of our heritage is to understand we are pieces of a larger puzzle. We're family members born into a unique and complex and rich tapestry of people.

When we choose to embrace our lineage, we choose to listen to the remarkable stories of the people who came before us and carry them with us like precious jewels. If we pay attention, our ancestors can become our true teachers. In their backgrounds we uncover a world of wisdom and wonder. Through the information we uncover through family tree research, DNA tests, and other traditional ancestry research, we gain a front-row seat to their lives as survivors. Their pasts can often create a road map for us to move toward the future.

Sometimes our research turns up delightful traits that have been passed down through our lineage—a musician or sportsman in the family that can empower us to move forward with our own talents and ambitions.

Sometimes we discover we've inherited certain things that were passed on to us that we're not even aware of—a smile, a voice, the color of our eyes or a cute little dimple, unspoken things that show up in the pictures and photo albums.

But sometimes the surprises seem surreal.

In interviewing families for this book, I've learned that most families have a strange secret or two: a mysterious death, an ancestor with a reputation. These secrets may not be life changing, but they certainly hit the jackpot for shock factor.

Through our searches we also often encounter serendipitous events and relationships—chance happenings that brought our ancestors together in unexpected ways. Or we might find celebrity cousins, ancestors who were imperial royalty, or relatives with a historical footprint or a famous bloodline.

Other times, our searches reveal the gift of a relative's unique stories that bring us closer to them, even though we are separated by decades and generations.

When we embrace the positive traits of our ancestors,

we gain insights that empower us to look at our lives from a broader historical perspective. Within these connections to those who came before us, we carry their souls and spirits as we prepare for the future.

### A Mirror Image: Her grandmother's legacy inspires genealogy career

*Mary Hall's interest in genealogy was piqued when she was just a child. Her family was eating out at a restaurant in town and she was reading the menu, which had a history of the town—founded in the 1650s—printed on one of the pages. "Hmm," she thought to herself, "the town founder's last name is the same as my mom's maiden name." Coincidence? Not so much. Her mom explained that they were descended from that guy from more than three hundred years ago. He was her something-great-grandfather.*

*Born and raised on the south shore of Long Island, Hall grew up immersed in the stories of her grandmother's family history. An amateur genealogist, Hall's grandmother audited college classes and scoured library files, pored through genealogy books, quizzed village historians, and contacted long-lost cousins to piece the family narrative together, long before the internet put centuries of information at our fingertips.*

When her grandmother created a family binder, Hall says, "I pored through it every day—names, dates, occupations, exotic far-off place names like Cork and Copenhagen and Gross-bockenheim."

"I learned I wasn't just half German and half Irish, but that I had long-standing colonial American ancestry, ancient English ancestors, and a Danish immigrant ancestor as well," she says.

At ten, instead of watching TV, Hall charted her extended family tree on poster board. Instead of playing video games, she entered family tree data into the genealogy computer program her uncle had bought for her. And with the advent of the internet, she realized that she could do genealogy research on her own.

She began researching her friends' trees and started offering her services to others as well—no one's tree is uninteresting or meant to lie undiscovered, she says.

"We all want to know where we come from—that's how we know how we got here and where we're going," she says.

Today, Hall is an online genealogist and blogger for the firm she founded, Heritage & Vino, and only just recently moved with her husband and two children from the New York village in which she grew up—the same village that was founded by her ancestor more than 350 years ago. She laughs that street signs bear the names of her cousins.

It is a career and a personal passion that she says she loves, and one that is driven by people's growing curiosity to find meaning in their lives. "It used to be that people contacted me because they wanted to find out if they were related to someone famous," she says. "I had to break some hearts."

Now, many of her clients are adoptees seeking information about their biological parents, and genetic tests are helping them find answers.

Others are estranged from their families, or their parents are estranged, and so they know little or nothing about their family history except what genetic tests are telling them.

Most significantly, she says, the desire is to feel more connected.

"People aren't as rooted anymore like they were in the times when everyone in town was related to everyone else in town," says Hall. "In my family, my brother lives in California; my husband's father is in Honduras."

Hall's beautifully written essay, included below, speaks volumes about how the gift of our *relatives' unique experiences impacts our lives. It is from her blog, www.heritageandvino.com.*

### Death Certificate for Hulda Wolbern (née Lindemann): Death Aboard the General Slocum Steamboat

Oh, I am so sad. I got the death certificate for Hulda Wolbern (née Lindemann) in the mail today. (Thank you, New York City Municipal Archives! I complained, but a two-month wait is not too bad . . .)

I knew this was going to be an emotional moment for me, possibly the MOST emotional moment for me as a genealogist, and I wasn't wrong. I think this is the most heartbreaking record I've ever had to look at. I'm not entirely sure why.

I've dealt with records for my own family and for clients where someone died young—in their twenties, leaving behind a wife and young kids, or even a kid themselves, never getting the chance to grow up and leave behind a legacy of their own.

But I think there's just something about the manner of Hulda's death, the personal family tragedy behind it, and the greater New York, German American, and American tragedy behind it. Hulda's death certificate tells me not just about the circumstances surrounding her death . . . it represents the deaths of all the women and children who died in the *General Slocum* steamboat disaster of June 15, 1904.

So, what does this death record say and look like?

It's from the Bronx, even though Hulda lived in Brooklyn. Most of the victims have Bronx death certificates because that was the borough closest to the disaster. The actual date of the certificate is June 21, 1904, because it took a few days for Hulda's body to be identified. Place of death: East River, off Port Morris (the southern tip of the Bronx, right across from Randall's Island). Character of premises (such as a home, hospital, etc.): Steamboat, "General Slocum." She was married, twenty-eight years old, born in Germany and living in the US and New York for eighteen years (which means I can look for an immigration record from around 1886). Her father was Caspar Lindemann. Her mother, for some reason, is not listed.

The certificate was filled out by Joseph I. Berry, Bronx borough coroner, and he identified Hulda's body in the morgue. The certificate says an inquest is pending—I wonder if that's particular to Hulda or to the *General Slocum* victims in general, if that inquest is public information, and if I would have the heart and stomach to ever read it if I could get my hands on it . . .

After his examination, Mr. Berry determined that Hulda's cause of death was asphyxia submersion. So Hulda didn't die in the fire—she drowned, as most of the victims did. She is buried in All Faiths Cemetery in Middle Village, as many of my German ancestors are.

On the second page of the report, we see that the

undertaker was R. Stutzmann. Rudolph Stutzmann was my great-great-grandfather. Hulda Wolbern was his wife's sister. Rudolph helped care for the remains of many of his (and therefore my) family members, but I wonder if it was particularly hard caring for his sister-in-law. I wonder if it was comforting to his wife, Augusta, and her parents and siblings, knowing that in the end their sister and daughter was in the hands of a loving family member.

Life is not endless. It's a journey with an off-ramp that everyone must take. When it happens, it's sad, but it's a fact of life. I've seen hundreds of death records. Everyone gets one eventually. But when it happens, we hope it's after a long, fulfilling life and that the manner is a peaceful one.

I can look at all these records somewhat objectively, usually, but Hulda's record is very emotional for me—because she was a young mother, because she lost her infant son, because they both died under such tragic circumstances. I couldn't find you for a long time, Hulda, but I hope you know you and your little boy aren't lost anymore. I found you. You are remembered.

## About the *General Slocum* Steamboat Disaster

The *General Slocum* steamboat disaster of June 15, 1904, left its mark on Mary Hall's family, but also on the lives of citizens of the greater New York area, and particularly the German American community. This tragedy struck the German Americans who lived on the Lower East Side hardest, as most of the passengers were from that neighborhood. Of the 1,358 passengers and crew, most were women and children, according to records published by the New York Public Library.[17]

"The aftermath of the sinking of the PS Slocum radically altered the German-American community of the Lower East Side forever," the New York Public Library article states.

The outing was supposed to be a day trip to get out of the city, but a fire started aboard the ship not long after it left the East River dock around 9:00 a.m. Though accounts differed, it was clear that the fire spread incredibly quickly. Since the boat was already out on the river, the passengers had no way of escaping the burning ship except to jump into the water. It's likely that many of the passengers didn't know how to swim, so they had to choose between the horror of the raging fire or the possibility of drowning. The type of clothing worn by women of the time also would have made swimming more difficult, but many of the passengers still jumped into the river.

Only 321 people survived the disaster. In the end, 1,021 people died in the span of a half hour, making this the highest death toll of any disaster in New York City history until the towers fell on September 11.

According to the New York Public Library article, "The shock of losing so many loved ones devastated families. Suicides and depression resulted from such a loss and many residents moved away." Though the German American community from the Lower East Side was hardest hit, since the ship had been chartered by a church in that neighborhood, Jewish, Italian, and other communities also saw the loss of loved ones from the fire. The entire New York City area was impacted.[18]

---

**Serendipity: Ancestry search reveals fate had a role in couple's pairing**

Joel Poznansky has spent his adulthood researching his family history.

He's British, the son of a Canadian and a Brit, and his wife is American, born and raised in Florida.

The couple *thought* they met by chance on the last day of graduate school when they were both invited to a party in Rhode Island. He was studying in Boston, and she was finishing her degree in New York City.

So, imagine their surprise when his ancestral research uncovered that his Polish Canadian paternal grandmother was buried a few feet from his wife's Floridian grandmother in a small cemetery south of Miami Beach.

The couple, who just celebrated their thirty-first anniversary, were there for the stone-setting ceremony for his wife's 102-year-old grandfather, whose family was one of the premier families of Miami Beach. His wife's grandfather was Alfred Stone, who moved to Miami Beach in

---

1929 with his wife, Lily, to build what was for many years the landmark hotel and tallest building in Miami Beach, the Blackstone Hotel.

"We thought we met by chance," says Poznansky, president of Wicked Uncle toys in Bethesda, Maryland. "It's possible that my paternal grandmother met my wife's paternal grandparents on the streets of Miami. It is strange, and fate."

He adds, "And to add further, my very English maternal grandmother, who I was very close to growing up, had told me a couple of times rather cryptically that a girl's ears were something to pay attention to. I told my wife that at one point as a joke, only to find out that she had been just as perplexed that her maternal grandmother (who died when she was quite young) had complimented her several times on her beautiful ears. We had both independently remembered the comments . . . because frankly, at the time, they seemed comical."

## Changing Places: "So who am I now?"

*With a surname like Von Driska, everyone always thought Sheila Von Driska was German, occasionally Dutch, maybe Austrian, but no matter what, it was a conversation piece.*

*"Von Driska . . . hmm . . . what is that?"*

*For fifty years, the San Francisco, California,*

*graphic designer found herself stating, "No. We're half Irish and half Czech." But not the Czech Republic of today . . . rather the Bohemia of yesteryear, the real Bohemia that used to be.*

*"I would always think . . . we're the true Bohemians . . . or at least half," she says. "And the fun part, Irish."*

*Until 23andMe.com came along.*

*"Just before I found out I was a mutt, I had gotten a dog from the shelter," she says. "My friend John called and said, 'There's a dachshund at the shelter.' I thought, no way. Sure enough, I went to the shelter, and there was a little six-week-old dachshund, and I got her and added her to my tribe of purebreds. But she grew into a dachshund on stilts, and her personality was the furthest thing from a stubborn German dachshund. She was playful, and I had to walk/run her five to ten miles a day because she had boundless energy. My nephew sent me the Wisdom Panel [DNA test for dogs], and we found out she was boxer/dachshund/mini pinscher. Aha. I was so grateful to know, and I read everything I could about boxers. I grew up with dachshunds my whole life, and she was so different. And suddenly I was in shape."*

That's when she decided to find out about her own DNA. The results: almost half British and Irish, about 30 percent Eastern European, and the rest French and German. She was surprised to see so much British, French, and German background in her family.

"My late aunt Marie would always tell us we had a castle in the Black Forest of Czechoslovakia and made me promise I would go look for it one day," says Von Driska. "She also argued with my father (her brother) that we were part Austrian, which led to her telling everyone that my ancestors lived very close to the Von Trapps. When my father married my mother, they started the dachshund lineage and named their first Baron."

Von Driska says she grew up with a belief that maybe she was a descendant of royalty. She says, "People would believe because of the 'Von' it could be true."

Her aunt left her a purple binder with possible instructions and a map called "Search for the Castle."

Her aunt had typed, "All of your grandmother Bess's people came from the same area of Bohemia: Velký Bor, Horazdovice, Blatná. Velký Bor and Blatná are about twenty kilometers from each other, or some sixteen miles. Relatively near is the birthplace of the first American male saint, St. John Neumann, b. March 28, 1811, Prachatice. Attended seminary in Bud jovice, contemporary with your great-great-great-grandparents. At least one-half of your inheritance on your grandfather's side is also Bohemian, since the name Jirik is a Bohemian one.

"Your great-great-grandmother Anastasia Jirik was born in Chicago, Illinois. Her parents came from Czechoslovakia. Your great-grandfather Frank Von Driska was also born in Chicago. Reportedly there were nine brothers, non-Catholic. Your great-grandfather became a Catholic when he married Anastasia Jirik. Because of the religious differences, early ties with the Von Driskas were not friendly.

"The couple was probably married in St. Procopius church. According to your great-uncle Bill Von Driska, who changed the original spelling, Vondriska, to Von Driska, your great-great-grandmother Von Driska went to the Church of God, he thinks. The story also goes that she came to Chicago along the RR tracks from somewhere else in the United States. I did not find any mention of Von Driska in the telephone directory in Prague, Pilsen, Bud jovice. The only Vondriska I found was Andelicka Vondriska, in Vienna," her aunt's note continued.

The castle story seems to come from a cousin of one of Von Driska's great-uncles, Bill, who met his cousin in recent years. It is not clear if this cousin spells his name Von Driska, Vondriska, or Wondriska. Her great-great-grandfather Bernard Babka studied a trade in Vienna. He became a plastering contractor in his own business.

Her aunt also gave her the names and addresses of the

great-great-great-grandparents on the maternal side, Joseph Killian and Anna Bartik, both born in Bohemia in 1839.

"That's as far as the search has gone," she says. "23andMe confirmed the possibility of the castle, so someday we may search for it. But now it will be exciting to figure out the British, German, and French parts of me that I never knew about," says Von Driska.

### LEGACY LESSON

"I guess the moral of the story is: I am a mutt and now I understand why I love the British sense of humor, French food and men, and of course my love for dachshunds . . . all things in my DNA. Who knew?"

### Tangled Roots: Digging up mysterious family legends

*Most everyone who does an internet ancestry search or DNA testing hopes to find an interesting story in their family tree.*

*But Laura Scott, an Austin, Texas, entrepreneur, found some real shockers. She learned quickly that researching your ancestry doesn't always turn up heroes.*

*First, she discovered she was descended from the brother of Oliver Cromwell, the English military and political leader who served as lord protector of the Commonwealth of England, Scotland, and Ireland until he died in 1653.*

> *But the real shocker: her grandfather Clar-*
> *ence McCullick is a descendant of Bartholomew*
> *Gedney, one of the Salem witch trial judges in*
> *Salem, Massachusetts.*
> *The finding stirred every emotion for her and*
> *her family members. "My mom was horrified, but*
> *overall, everyone thinks it's a pretty unique his-*
> *tory," she says. "My kids think it's kind of cool*
> *and they have told all their friends."*

Her research uncovered the ancestor trail back to the 1700s—many from Ireland were drafted to fight in the American Revolution.

But the connection to Gedney was the most intriguing and surprising.

She learned that Bartholomew's father, John Gedney, was one of Salem's founders and leading citizens, and Bartholomew followed in his father's footsteps, both of them highly respected in the community. As a settler of the town, he was involved in the local military and rose to the rank of colonel. He also helped grow the town's economy as one of the town's original builders and was responsible for the town's gristmill and two sawmills.

A true story of the good and the bad, Scott says the findings caused her to think twice about when she makes assumptions in her life now.

"I guess I thought that the Salem witch trials were

gruesome and horrible," she says. "But once I looked at through the eyes of the early evangelical movement in the US and [thought about] how hard this tradesman/ merchant/smithy/religious leader/carpenter/physician had worked to build the town of Salem, I understood differently. I could understand how someone would have to serve as a judge because he was a town elder, and not necessarily want to do it, but have to stand in judgment with what his cultural mores and the times were back then."

"Understanding what he did to build the community caused me to have empathy for him in having to deal with what was going on at the time," says Scott.

Scott, who is the owner and cofounder of Mosaic Weighted Blankets, says uncovering this piece of her ancestry changed the way she looks at life.

"I think it's changed my attitude about myself," she says. "I used to think I was nobody important. But seeing that I am tied to something greater, these people in our history, makes me feel like I'm part of something bigger.

"It makes me feel more entitled to be here, and I can see why many are so fascinated by the witch trials, which are still very much a part of the American identity."

## LEGACY LESSON

"I'd encourage everybody to investigate their ancestors' history. The story is there, whether you want to know it or not. But there might be an endearing legacy there just waiting for you to claim it. You must do the work if you want to know that legacy. In a way it feels like I'm on an episode of the TV show *Who Do You Think You Are?*"

### The Lookalike: DNA reveals her father's link to one of North America's most notable skeletons

*About seven years ago, when Alexis Sánchez was checking out books at her local library in Kennewick, Washington, she spotted a bronze bust that had a startling resemblance to her father, Javier.*

*She remembers being immediately drawn to the glass case where it was perched and thinking, "Why do they have a statue of my dad?"*

*"I just had this weird connection or feeling," says Sánchez, thirty-three, who now lives in Austin, Texas.*

*The bust was a sculpture of the Kennewick Man, called the "Ancient One" by Native Americans, which experts say is one of the oldest skeletons ever found in North America. The nine-thousand-year-old remains were discovered in the 1990s along the Columbia River by two spectators watching the annual hydroplane races. Through DNA analysis, the remains have been traced back to multiple Native American tribes that lived along the Columbia River in parts of Washington and Oregon.[19]*

*"I'd always had this fascination with Native Americans, and for some reason identified with them, even though my dad was an immigrant from Mexico and my mom had British and Irish roots," she says. "I was the kid who pretended I was Disney's Pocahontas" (the animated Native American princess).*

*On her way home, she called her dad and said, "Dad, you won't believe how much you look like the Kennewick Man statue," she says. "He just laughed and shrugged it off. But something in me felt we had a Native American connection."*

*When she got home from the library, she posted on her Instagram page about the coincidence. But she says she always had a nagging feeling about it.*

Until 2019, when Sánchez received the results of her 23andMe.com DNA test, and the tests she'd coaxed Javier and mother, Lorie, to take. Her mom's ancestry was no surprise—the results revealed she was 100 percent European (Irish and English).

Her fifty-nine-year-old father had fled Mexico when he was seventeen and eventually moved to the Tri-Cities area in the state of Washington. But, after seeing the Kennewick Man statue, part of her hoped she'd discover he was part Native American.

In the end, not only was Javier part Native American, Sánchez discovered that she and her father were linked directly to the Kennewick Man along his paternal line.

"I was like, 'Wow, that is super cool!'" says Sánchez, a former nurse who is studying to be a musician. She is the oldest of three children, with siblings Alisha, twenty-five, and André, twenty-seven.

She called her father, who trekked over to the local museum where the bust now lives to take a look at the statue himself.

"There's some resemblance, but I really don't think I look like him," says Javier, a legal permanent resident who is seeking his American citizenship. The news spread around town, and he became somewhat of a local celebrity when a story was published in the community newspaper. "Everywhere I went people would kind of stare at me, kind of sizing me up to see if I really did look like him. It was the talk of the town for a while."

For his daughter, the DNA results "tied a pretty little bow" around the fond feelings that she'd always had toward Native Americans. The DNA results also included links to the Aztec Indians and a sprinkle of French.

"I just always assumed my ancestors were Mexican, but this makes sense if you study the history of Mexico," says Javier. "I really at first thought this whole DNA testing was hocus-pocus, and the test sat in my home for a while after my daughter sent it to me. But to think they can trace your DNA to nine thousand years ago, that is something."

Both father and daughter still can't get over the coincidence that Sánchez lives in the same area in America that the Kennewick Man hailed from (or at least where he was buried), which is about six miles from where Javier lives.

"My dad was escaping a bad situation and left with the clothes on his back," says Sánchez. "It was really dangerous in the border towns. He had to turn back from the border twice before he found a way to get across. He hopped on trains and freights and found himself in the bitter cold in Nevada, in the middle of winter, then on to Portland and finally to Washington. He came here for a better life. He started loading potato trucks, where he met my mom, who was a truck driver, and he slowly built a life for all of us."

Today he owns his own shipping pallet business and has built a beautiful house with marble floors on a large piece of land.

"I told him I kind of thought he overdid it with his marble floors and countertops, and I would tease him," she says. "But he said he grew up with dirt floors and not having shoes and that he can have marble floors if he wants."

"It always bothered me that Native Americans live on this side of the border and Mexicans live on the other and aren't exactly welcomed moving here," she says. "Since discovering my father and our family has Native American roots, it really [demonstrates] how all of us are the same people and how there should be no wall to separate us. That is our heritage—we are more alike than different. In our family it took a DNA test to prove that."

## LEGACY LESSON

"In the end, DNA testing is showing that we are all related, regardless of what side of the border you come from. We are all human beings who come from the same place. We should end our divides in this country and remember we are all in this together."

### My Grandma, Myself: Family secrets lead to lessons in grit and imperfect resilience

*For more than ten years, Don Grossnickle has dedicated his life to helping high school football players who have been struck down by paralyzing spinal cord injuries navigate the huge challenges they and their families face as they try to move on in life with their serious injuries long after the game is over.*

*As cofounder of Gridiron Alliance, an organization dedicated to promoting awareness and prevention of catastrophic injuries to high school student athletes, and a deacon in the Catholic Church, Grossnickle preaches the power of grit and resilience to get through life's toughest* times and to move forward. Though he retired in 2003, he still consults with the two paralyzed athletes who cofounded the organization with him. These days he is involved in various volunteer organizations, including work in Uganda, East Africa, helping mothers and babies suffering from malaria.

At much the same time that he was cofounding Gridiron Alliance, Grossnickle launched his own personal crusade to explore the lives of his ancestors, especially his Swedish grandmother, Maja. Maja was his mentor and the woman he aspired to emulate, and he had always been in awe of the stories she told about coming to America and surviving her husband's sudden death and the Great Depression.

For Grossnickle, his grandmother epitomized the lessons in perseverance he was trying to inspire the athletes with.

Through his research, he discovered his grandmother's steamer trunk that she'd brought with her from Sweden as an immigrant to America in 1922. Inside he found photos, archives, and a treasure trove of family history. As a child, he'd always been drawn to the legacy stories she shared and had savored the special mentoring relationship he had with his grandmother.

*When he found this trunk, he stumbled on a few family secrets that rocked the reality he thought he knew.*

"My sweet grandmother, Maja (Marie) Kallgren Wittenstrom, was just thirty-five years old in 1932 when she found her husband dead on the dining room floor," says Grossnickle. With her three children, ages seven, five, and one at the time, she was forced to carry on in the face of America's stock market crash that killed her husband's start-up auto business and most likely contributed to his heart attack, devastating the young family's Swedish American dreams.

Almost overnight, his Swedish immigrant homemaker grandmother became an unwilling grieving widow alone in a hostile land without a job, he says.

"Maja was forced to pick herself up like every Humpty Dumpty–like survivor must do," he says. "There would be no one coming to her rescue. Maja had to discover the limits of her grit and resilience, guts and gumption, as she fought back, or [she would have been] further crushed in emotional defeat."

Doing whatever she could to survive and thrive, she opened a boardinghouse and became a night nurse for wealthy families.

"With three mouths to feed, Grandma Maja dug deep, working the night shift, coming home to prepare the children for school and care for the boarders," he says. "She became exhausted and dealt with chronic homesick moments. Maja called upon childhood experiences and schooling growing up in Sweden, her professional nurse's training, and her Swedish ancestors' tales of resilience in action, which inspired her spirit. Through it all, Maja was driven forward to keep sight of her goal to one day see her children become self-reliant and capable."

During his digging into his family's past, Grossnickle saw time and time again the worthiness of his grandmother's heroism. But that's when he also uncovered several family secrets that had been buried in the forgotten past.

He says, "After eight years of stress and strain, Maja, in 1940 buckled under the pressure. Her kids were temporarily farmed out to makeshift homes as she was voluntarily hospitalized. Maja's spirits sank low in depression, and shock treatments and crude experimental methods somehow helped heal and restore her to health, and her family was reunited."

Another family secret came to light, sadly revealing that in 1949, likely owing to overwhelming guilt about failing to rally sufficient grit, his grandmother attempted to take her own life. Thankfully, she recovered to spend another twenty wonderful and happy years with her family and grandson Grossnickle.

Uncovering the "secrets," Grossnickle realized the lessons in fortitude and survival unearthed by this new knowledge of his grandmother's trials painted an even more powerful lesson for his life—one he could share with the young athletes he was working with. To survive, each of them had to dig deep daily and find reservoirs of strength, perseverance, and fortitude.

"Every grandson should have the benefit of spirited role models like Grandma Maja to inspire and guide personal development," he says. "In the end, Grandma Maja masterfully taught problem-solving skills—exhibiting brave courage toward keeping one's balance while untangling the logjams one willingly sets out to conquer. Maja called upon her faith in herself, her children, and a close kinship with God. Each of us might be fortunate to take opportunities to pass along stories with our very own grandkids, reflecting on our own imperfect life experiences.

"Maja found a way to compassionately forgive and heal her past and kindly left behind an oral history legacy that now can be shared more widely so that future generations might have the benefit of studying her use of personal resilience to address their unique challenges," he says.

## LEGACY LESSON

"My grandmother beautifully portrayed how our ancestors and their inspirational and sometimes imperfect stories contribute to the future. With twenty-twenty hindsight as a gift, it is our job to share their legacy with others. Grandma Maja first inspired me to become a teacher. With advanced degrees, I wrote articles and books and spoke about the problem of high school dropout rates. [My work often] focused on legacy lessons learned, which I shared with parents and teachers all across the nation. Grandma Maja's legacy inspires my work today as a medical missionary working to save lives of mothers and babies with complicated malaria in Uganda."

## CHAPTER FIVE

## BACK TO YOUR HEALTH FUTURE:
### It's in your genes

One of the benefits of tracing your family history is it can help predict the future if you use information from the past.

This is especially true when it comes to medical matters. You can't change your genes, but with insights into your DNA medical makeup, you could learn about what conditions or diseases you might be susceptible to or that you have an increased risk of developing. These findings can ultimately make you healthier if you follow up on those results and get tested regularly for the medical red flags that have shown up.

It's important that you not make assumptions about

your results, as there's room for misinterpretation. If there is genetic medical information you're concerned about, it's best to have it reviewed by your physician, who will refer you to a genetic specialist if it seems further testing needs to be done.

The efforts of the people in this chapter to learn more about their health and heritage reveal the importance, and sometimes challenges, of understanding one's medical background.

### Matters of the Heart: Uncovering family patterns to get and stay healthy

*When Lynell Ross was in her late twenties, just after she got married, she discovered that she had unusually high cholesterol.*

*"It upset me, and I did some research on how to lower it, but back in the eighties there wasn't much talk about the link to high cholesterol and family history, so it went unnoticed," says the Auburn, California, mother of two grown children. "A few years later, after I had given birth to my second son, my mother had a heart attack at age sixty-two. She lived a few months but then had a stroke and passed away. We were all in shock and just tried to pick up our lives and go on the best we could, my two sisters, my husband, and my sons."*

*Over the years, she began exploring her family history to uncover health and lifestyle patterns in her relatives on both sides of her family. Ross's efforts to learn about her health and her heritage revealed the importance of looking into her medical background.*

*Her journey has led to a trailblazing career as a certified health and wellness coach, behavior change specialist, and nutrition and fitness trainer, helping to heal people's lives. She is also the editor of Zivadream.com.*

Delving into her family heritage and health history was a tedious and relentless task. She found information on her mother's family running in the Oklahoma land rush in a box of papers she'd kept from her mother. She had pictures, deeds, and documents dating back to the 1800s.

"I made scrapbooks of the documents, and my younger son even used some of them in school reports," she says.

With so many of her family members deceased, she faced many challenges collecting the health information. In addition to old-fashioned research, she conducted interviews with her mother's best friend and a woman who had been married to her cousin (her cousin had also died of a heart attack when he was sixty).

Through the digging, she discovered that both of her grandfathers had passed at young ages—forty and fifty—

from heart attacks. Her father's mother passed away from a heart attack at age fifty-five. Her mother's mother made it to age seventy-two, but only after going through two operations related to heart disease. Then her cousins on both sides began to die from heart attacks or suffer from strokes in their fifties.

"No one had noticed that heart disease was taking our family," she says.

It inspired her to go back to school and study nutrition. "I wanted to prevent heart disease from hurting any more people in my family," she says. "Within my first semester back at college, I learned that heart disease is the number-one killer in America. Before this, I had been in marketing, living a fast-paced life, working full-time in addition to being a wife and mother of two active sons. I began to worry about my heart."

Armed with her family history of heart disease, Ross made an appointment with her doctor, who confirmed the fears her findings had surfaced: heart disease can be inherited.

"My nutrition teacher helped me to make an eating plan for myself and my family," says Ross. "I found the information on nutrition and fitness so important that I decided to become certified as a nutrition and fitness trainer to help others. I kept researching about health, and the more I learned, the more compelled I felt to share this life-saving information. That's when I decided to write a book, to use as a guide with my clients and in teaching workshops."

Driving her research was her goal to stop the cycle of heart disease in her family.

"One of the reasons I did my research on what it takes to live an optimally healthy life was to help prevent and manage heart disease for my husband, our children, my sisters, and my nieces and nephews because of what happened in my family," she says. "On my journey, I learned that great health doesn't just come from what we eat or how much we exercise, but also how we manage stress and connect with other people. The way we think and our perception of stress can add to heart disease, type 2 diabetes, and even certain cancers."

Through her own family she saw the toll stress had taken, possibly contributing to the trail of heart disease.

"My family on both sides were very hardworking people," she says. "I found that my great-grandparents came across the United States in the Oklahoma land rush. My grandmother owned a restaurant in Oklahoma in the 1920s, and after her husband died, she went to work in the shipyards in Oakland, California, during WWII.

"My other grandmother owned a car dealership in San Francisco in the 1950s. The women in my family were as strong as the men. They did what they had to do to survive, but what was passed down to me, along with a hard work ethic, was the inability to slow down and rest. My research on health and wellness showed that we need to rest and take the pressure off ourselves to manage chronic illness like heart disease," she adds.

In addition to the genetic link to disease and health conditions, Ross believes that the lifestyles and work ethics of our ancestors play a direct role in how we manage our stress and health risks.

"I often wonder if today's baby boomers are under so much stress because of the pressure we put on ourselves, that has been handed down to us," she says. "I encourage my clients to be as self-aware as possible, and this means understanding where our parents, grandparents, and ancestors came from and how they lived. Research proves that we carry within us family patterns, such as a fear of not having enough or not doing enough. So we push ourselves to work harder and keep going, leading to burnout and illness."

As she learned more about tools to help her slow down, such as yoga, meditation, and mindfulness, Ross says she believes that those on the spiritual path are among the most healthy and happy people.

## LEGACY LESSON

"One of my mentors says that we teach what we most need to learn. So, I continue to learn, to practice what I teach, slowing down, taking pressure off myself and encouraging others to do the same. I am now the age my mother was when she died—my youngest son was just one year old [at the time]. My children grew up without grandparents. We just found out that our son and his wife are expecting their first child. I plan to do everything I can to take care of myself and my husband so we can be there for our grandchild as long as possible."

## Tracking Breast Cancer and Ancestry, With Mysteries in Each

*In 2011, Amy Byer Shainman embarked on her ancestry search with two missions in mind: to study her family lineage and research more about her family medical history. She was attempting to find out more about the BRCA gene mutation she inherited that put her at high risk of developing certain cancers.*

*Shainman's older sister had been diagnosed with three separate primary cancers—ovarian, uterine, and breast cancer. Through her experience making decisions about her health, Shainman became invigorated with the purpose of establishing herself as a leading advocate for those with BRCA and other hereditary cancer syndromes.*

*Today, Shainman tirelessly works to educate others facing the same daunting reality. Known as the BRCA Responder, she is a patient advocate who provides support and education*

*surrounding BRCA and other hereditary cancer*
*syndromes. Certified genetic counseling is crucial*
*in the genetic testing equation, she says.*

Shainman, now fifty and a mom of two, had no red flags for developing cancer back in 2008. But after her sister tested positive for a BRCA1 gene mutation, Shainman met with a certified genetic counselor in 2009 and discovered that she too carried the same BRCA1 gene mutation that put her at high risk of developing breast and ovarian cancer. Shainman and her sister's specific BRCA1 mutation is one of the three founder mutations associated with those of Ashkenazi Jewish ancestry.

"My sister had three separate primary cancers and survived," says Shainman. "I wanted to know more about my family medical history and ancestry, so I delved into researching family history on my dad's side. His mother (my grandmother) passed away at thirty-three from metastatic breast cancer, my great-grandmother had a cancer of her ovaries and died at forty-six, my great-grandfather died suffering from rapid weight loss (which could be attributed to pancreatic cancer), and a few of my grandmother's siblings had colon, melanoma, and pancreatic cancers. The family history of cancer was profound."

Shainman's children were eight and five at the time when she decided to undergo two prophylactic procedures—a complete hysterectomy and a double mastectomy.

"And then I just went on with my life," she says.

For the last decade, however, she has become one of the nation's most passionate advocates for the prevention of hereditary cancers. She gives speeches, and she blogs about it on her website, thebrcaresponder.blogspot.com.

"For a lot of people, this history falls through the cracks," she says. "If my sister had not been diagnosed, it could have remained hidden. But instead it made me become vigilant about my health and also proactive; I chose to have preventative surgeries to drastically reduce my cancer risk."

That is why Shainman has turned advocacy into her life's work. "I couldn't remain silent," she says. "My options seemed to be shut up or save lives."

The West Palm Beach, Florida, mom documented her journey in her book, *Resurrection Lily: The BRCA Gene, Hereditary Cancer & Lifesaving Whispers From the Grandmother I Never Knew.*

## LEGACY LESSON

"Important decisions regarding one's health can be based on genetic test results. Whether it is regarding your health screenings or surgical decisions, you want to make sure the tests are accurate," she says. "The most qualified individual to assess one's cancer risk, order genetic tests (if any), and interpret results is a certified genetic counselor. A certified genetic counselor has specialized training in medical genetics and counseling. They know the ins and outs of insurance. They know the most trustworthy labs. They are most up-to-date on new emerging gene mutations. Genetic counseling is a conversation. A certified genetic counselor can help you understand what genetic testing may mean for you and your family. Speaking with or meeting with a certified genetic counselor does not mean that you have to undergo any type of genetic testing. If you feel your genes may be putting you at increased risk of developing certain cancers or have concerns about other genetic-based health risks, speak to a certified genetic counselor by phone or in person. These resources can be found through the National Society of Genetic Counselors."

---

### Discovering a History of Mental Illness: Here's How

*The following section is excerpted from a post by Becky Klein McCreary, newsletter editor for the* Southern Arizona Genealogy Society *in Green Valley, Arizona.*

Learning that an ancestor spent time in an institution can be troubling and may be a roadblock in your genealogy research. Some records may be viewed, others are private, and many have been destroyed.

There may be a story about a family member who was mentally or physically impaired, imprisoned, or who died in a poorhouse. You want to know more about him or her.

Or you might find hints on a special census. For example, in a move reflecting the prejudices of the time, the US Federal Census—1880 Schedules of Defective, Dependent, and Delinquent Classes collected information using the following categories: insane inhabitants, idiots, deaf-mutes, blind inhabitants, homeless children (in institutions), the imprisoned, and paupers and indigent inhabitants. There may be in some states an attachment detailing the disabled person's condition.

As American communities grew in the nineteenth century, institutions were established. [Before that time], families took care of those who needed help.

Begin the search by putting the person's name into search on Ancestry.com or FamilySearch.org. Also look at census records for the town in which the institution was located. There can be, as with all records, scant or incorrect information.

If you have the person's name and the location of the institution, read the history of the facility. When was it in operation; how did it operate; how was it funded (state, county, religious, private, etc.); and does it still exist?

If it was state or county funded, records may be at the county auditor's office or they should have knowledge of where the documents are stored, if they were kept. If the institution still exists but with a different name, the website will give their research policy.

Browsing online, you can find databases and indexes

to state archives and historical groups. The National Union Catalog of Manuscript Collections (NUCMC) includes manuscript collections, archives, and oral history collections from throughout the United States, and other materials are held by eligible archival and manuscript repositories throughout the United States.

Each state, county, and church has its protocol for privacy. Expect to be told that records don't exist.

Hospital records for the deceased are seldom retained after several years, but it doesn't hurt to try. Insurance companies are commercial businesses, so there is no obligation for them to keep old records. But again, it doesn't hurt to try.

I tell researchers to never be ashamed or embarrassed by the adverse events or secrets they find about an ancestor or relative. It may give [them] a better understanding of the person and his or her family, and [it] is a view of the social history of the time.

## CHAPTER SIX

## WALKING IN THE FOOTSTEPS OF YOUR ANCESTORS:
### Is geography in our bones?

With the popularity of DNA tests helping people find out more about our ancestors' origins, many of us are driven to travel to our ancestors' homelands and fulfill a dream to literally walk in the footsteps of those who came before us.

This interest has spawned a boom in the heritage travel trend as we embark on life-changing journeys to the places of our family's past.

For many of us whose family immigrated to the US, visiting our ancestral home feels so familiar, as if we are being pulled to an overseas place where we will connect

with our past and find a sense of wholeness within ourselves. What makes us feel an intense sense of home when
we walk through streets we've never walked, recognize
faces we've never known, and experience inexplicable connections?

Returning home, we carry the place back with us, as
if we've packed our roots in our suitcases and brought the
transformative trips home in our hearts forever.

It can be a very emotional and humbling experience.
We may have discovered why our ancestors left their country to begin a new life in a foreign land. We may have
begun to understand deep down what we have inherited
from our families' cultures of origin.

### The Accidental Tourist: Woman reunited with long-lost cousins in Sicily

*When Allison Scola was backpacking through
Europe after college, a tray of Sicilian pastries
delivered to her in Rome unlocked an entire world.*

*Scola's grandfather Vincenzo immigrated
to the US from Porticello, a tiny fishing village
in Sicily, in 1913. In the late 1920s, he traveled
home to the largest island in the Mediterranean
to marry, then returned to Brooklyn, New York,
with his bride and started a family. Scola's father
was born during World War II, and although he
had traveled to Italy—even living there for a year*

in the 1960s—he didn't share much about his Sicilian cousins or family with his daughter.

In 1996, when she was twenty-four, Scola was inspired to visit Sicily after meeting her father's first cousin Pietro, a pastry chef, in Rome. His cannoli—a dessert she had delighted in with her father during her entire life—were like nothing she had ever tasted in the States. Pietro convinced her to travel to Sicily.

There, the singer and songwriter met a second cousin, six months her junior. It was kismet—the duo felt an immediate organic and spiritual connection.

"I immediately thought, 'this is where I belong,'" says Scola, who now lives in Englewood, New Jersey. "I felt so at home in Sicily. Arriving there for the first time was overwhelming."

Since then, she has become a frequent tourist. "Every time I get off the plane, I feel so good; the energy just takes over," she says. "I'm always astounded by the feeling of connectedness I have to the landscape and to the people. I feel like it is where I belong."

In 2013, Scola and her Italian cousin Evelina cofounded Experience Sicily, a boutique tour company. Together they plan beyond-the-ordinary itineraries in Italy, many that serve

*Italian Americans conducting heritage research. The team not only collects vital records at town registry offices in Sicily, but using their deep network of local insiders, when possible, they reconnect American families with living distant relatives and plan authentic experiences so their clients truly understand the history, culture, and land from which their ancestors emigrated.*

*Even though she'd formed a deep connection with her dad's mother's side of the family, a mystery still remained about her grandfather's family. Her father knew very little, if anything, about his living relatives.*

*Her journey to discover her father's paternal roots began.*

Through Experience Sicily, Scola shares her passion for her family's ancestral homeland with other Americans seeking to connect with their Italian roots. By planning ancestral pilgrimages, she helps fellow Americans tap into the same connectedness and rootedness she feels every time her plane hits the tarmac in Italy.

In anticipation of their clients' travels, Scola and her colleague (and husband) Joe Ravo spend hours online scouring vital records to unpack family histories. Their expertise at locating and examining timeworn documents enables them to find forgotten relatives and piece together personal lineages. "It's so rewarding when we connect distant cousins to distant cousins," she explains.

Her father's longing to connect to his paternal relatives took a heightened turn in June 2019, when age began catching up to him. Scola decided to turn the tables on herself and make a concerted effort to find the family he was seeking.

While staying in Porticello, the village of her grandfather's birth, a tip from a new tourism contact laid the bread crumbs. By recommendation, Scola introduced herself to the owner of a local restaurant who, she was told, might be a connection.

After dining at the trattoria, she introduced herself to the owner—Sebastiano Scola—who was equally fascinated to meet her. After talking for a few minutes about family history, they determined that their great-great-grandfathers were probably brothers.

"He explained that there are only a handful of Scolas left in Porticello, and knowing the houses that my great-grandfather built, our connection was clear," she says.

Armed with this new information, Scola set off to the office of vital records in the town of Santa Flavia. There she spent a couple of hours with a very patient director, who located the documents that supported what she had already found on Ancestry.com. "From the meeting, I was able to confirm my family's history as founding members of the village and, importantly, glean that I come from centuries of professional fishermen and fishmongers," Scola says.

Returning to Sebastiano's restaurant with more details, she spent a couple of fascinating hours with Nino Scola, a distant cousin, and his wife and daughter, who Scola says has many of the same characteristics as she does. "When you see us next to each other," she says, "it's clear we're related."

In September 2019, Scola's father returned to Sicily for a visit. Armed with specifics about his father Vincenzo's side of the family, he met Nino Scola and Sebastiano's father—his contemporaries. The men spent an entire afternoon discussing their life stories while gazing at the sea where their grandparents and great-grandparents had fished for tuna, anchovies, and sardines for centuries. "He has been to Sicily more than a dozen times," Scola explains. "But this time my dad was even more emotional when he returned. Connecting with these distant cousins gave him a new perspective on his father's legacy."

## LEGACY LESSON

Serious ancestry seekers should embark on a journey to their family's homeland.

"To truly uncover your family history, you have to go. Do some research before you leave. Create a family tree using the tools available on the internet. Collect information from your oldest living relatives—record even the most convoluted stories. You'll see these will reveal truths. Then, make the time to go and talk to people in the town. Arrange to have an English-speaking guide or local insider with you. Have patience. These tiny villages have few staff members and limited office hours, so plan accordingly. Stay in the village or town and spend some time. That's how you discover your personal history, and when you do, it will change your whole perspective on your life."

### Time Traveling: "I found a part of myself I didn't know was missing."

*Kalev Rudolph, twenty-four, grew up in south-eastern Philadelphia, but from an early age, their parents inspired them to explore different cultures, especially Estonia, from which their family emigrated more than four thousand miles to America. Their household dinner table talks were flush with stories of the family's history, adventures, and tribulations during World War II and their grandparents' love story.*

*The need to know more about the specifics of who their grandparents and ancestors before them were, where they lived, or what they did is insatiable for Rudolph, who now lives in Indonesia and makes their living as a travel blogger, foodie, and writer.*

*DNA testing gave them the tools to take their*

*exploration to an even deeper level, Rudolph says.*

*"Learning about your family history is transformative," says Rudolph. "When pieces of the historical puzzle that have led to you come together—something changes. As a second-generation Estonian American, I have been lucky enough to know pieces of my grandfather's coming-to-America story but never thought I'd have to do much searching for answers."*

Rudolph's yearning to explore that rich history started shortly after their eighteenth birthday, when their grandparents took them to Estonia.

"I learned things I would never have thought to ask," says Rudolph.

"Finding out how super-Estonian my grandfather was was shocking," says Rudolph. "While he has some crosses from other countries, he is a massive majority of Estonian blood going back for generations."

Fortunately for Rudolph and their cousins, their grandparents were determined to take their heirs on these exhilarating journeys back in time to soak in their family heritage.

After the USSR fell and Estonia's borders opened again to US citizens in 1991, their grandparents returned to their homeland as many times as they could. They had originally left in 1944. It became something of a family tradition to take whichever grandchild had just turned eighteen to Estonia in the summer to get a tour of the family history.

"All the while, letting slip family stories and bickering over maps—we were often lost for hours. After our first two weeks of meeting cousins and enjoying the nearly twenty-four hours of Baltic summer sunshine, we made our plans to visit the land where my grandfather's (Vana-isa, as we called him) childhood farm was, as well as the grave of my great-grandfather."

It's a tiny town on the northern side of the country, speckled with crumbling medieval castles, simple farmhouses, rolled bales of hay (often with herons perched on top), and the brutal vestiges of Soviet occupation.

For Rudolph, the journey was magical.

"The plot where their farm was had become a glowingly green field, just waves and waves of sunlight rippling across the tall grass," says Rudolph. "All my life, I had understood this place as a symbol and a story—but not somewhere real."

The experience was overwhelming and heartwarming all at once.

"Standing there, next to my grandfather—looking at the actual earth where he played and laughed and would eventually hide in the woods for weeks avoiding the Soviet soldiers, eating grubs and plants for days—felt indescribable," says Rudolph. "It was beautiful and quiet, and full."

Rudolph got to envision the stories they had always been told and describes it here:

"For his sixteenth birthday, my grandfather was supposed to get biplane lessons," he says. "As a well-to-do family, they had a housekeeper, outdoor 'fridge,' house, and plenty of land. When news of the Soviet invasion came, they buried what silver and valuables they could under the barn, took a fully loaded pistol and a few prized possessions, and fled.

"The pistol was to be used by my great-grandfather if the soldiers came too close, first killing my grandfather, then his mother, and then himself. This moment never came. But there were close calls. They spent weeks in the forests around their home, foraging what they could and hiding day and night.

"When my grandfather's birthday came around, their housekeeper, who was kept on by the now occupying Soviets, snuck a pancake out to their family. It was his only gift.

"Working slowly toward the coastline, they moved during quiet hours, doing what they could to stay unnoticed. After weeks, they made it to the intended meeting point, a small dock on the western shore of the country, where a boat was waiting to take them across the Baltic Sea to the small island of Saaremaa. Once there, they organized transit into Germany, where they'd stay for the next two years dodging the draft and trying to find transport to the US.

"On my trip to Estonia, more than fifty years later, we spent a long day trying to find this dock. My grandparents mutually decided to skip this visit, so it was just my mom, stepdad, and me. We drove through town after town, using the churches as guideposts, stopping along the way to ask strangers directions.

"We'd been given handwritten directions, based on my grandfather's best memory, along with some town names to get us going the right way. Finally, we found a pastor who was able to point us down a road unlisted on our phone maps. I remember how long it was, driving incredibly slowly in our rented Peugeot over rough gravel and skull-size rocks.

"The trees were tall and thin and bleached white. Wind rustled the birch leaves, and we all made jokes about driving to our deaths. And we held silence in the comfort of our air-conditioned car, looking down the same road my grandfather and great-grandparents traveled fearing for their lives.

"At the end of the winding drive, we came upon the collection of old buildings, which were eerily empty, one smokestack chugging along. No one came out to bother us or ask why we had come, so we found a place to park. My mom got out, having been once before, and led us to the small concrete dock leading straight into that cold northern sea.

"It's a small bay. On that summer afternoon, with waves lapping at the moss-covered posts, it couldn't have been more tranquil. We stood and stared into the water. I couldn't help but imagine the small pontoon boat, the tension that must have hung in the air. I wondered what they whispered to each other.

"My mom told me they made it there just before dawn. They rode away from the eastern sunrise, into dark morning waters and away from home."

The trip opened a door to a whole new sense of self, says Rudolph.

"Instantly, I found some part of myself I didn't know I was missing," says Rudolph. "It was something I maybe knew existed but had never known. And not only a part of myself, but part of the man who raised my mother. These were pieces of the trauma and perseverance that define my family."

Rudolph adds, "Gaining this knowledge gave me the tools to better connect to myself and my history. It helps me make peace with and find a fuller appreciation of the life I have.

"Traveling to learn about my ancestor's homeland has inspired me to want to see as much of the world as I can," says Rudolph.

## LEGACY LESSON

"While I don't necessarily think of my work as finding a voice for my family (they have been speaking for themselves for generations), advocating for my community and other groups around the world is central to everything I do—tenets I feel are deeply connected to my history and ancestors," says Rudolph.

### Getting Back to his Roots: "I felt like I came home."

*In 2002, Anthony Bianco found himself standing in Friuli Venezia Giulia, a northeast Italian region bordering Austria, Slovenia, and the Adriatic Sea, where his father grew up.*

*"I felt like I came home," says Bianco, forty-six, a father of two daughters from Brisbane, Australia, and author of the blog The Travel Tart—Offbeat Tales From a Travel Addict.*

*Bianco, whose parents both emigrated to Australia as children in the early 1950s following World War II, was backpacking around the world and had decided to see where his parents came from—his father's side in Friuli, and his mother's from Sicily. His parents met as adults in Australia.*

*"My grandparents left all of their family behind back then—and on my mother's side, they never set foot in the country again," says Bianco. "But I was able to meet some of their brothers and*

sisters (such as my great-uncle, who was a cobbler in Sicily well into his late eighties!) on a trip before my grandmother passed away, so I was able to make connections from her stories to the physical landmarks and real people." For the first time, Bianco says, he could emotionally place himself in his grandparents' lives by meeting his relatives.

It began an almost twenty-year pursuit to learn more about his roots.

"I wanted to explore what it must have been like to move halfway around the world to a place that you don't know at all and sometimes just had heard of," says Bianco. "My family's story is about how they decided to make the move because they wanted something better. It was a totally different era that most of us today would find totally alien, but this is where I came from—very humble origins."

He also feels compelled to trace the history to pass it on for generations to come. "I think it's important that my daughters know the background to their heritage," he says.

His mother, Maria, chronicled her experience leaving Sicily and starting a new life in Australia in a book called Three Trunks and a Cardboard Case (named appropriately for all the worldly belongings the family was able to take with them).

*In the book, his mother describes how her father, Bianco's grandfather Salvatore, leaves his native Sicily forever for Australia by himself, never to return. One year later, his wife, Nunzia, and children, Vito and herself, leave Sicily as well, and they embrace a new life in the isolated and humid sugarcane fields of far north Queensland, hoping for a better life.*

*Raised in Australia, a home of tranquility and tropical beauty, Bianco feels compelled to continue the next chapter for his mother and to find out more about where both his mother and father came from, in order to connect with something that is exclusively his family's and personally enriching for himself.*

"My family on both sides were typical of what happened to many families in post–World War II Italy," says Bianco. "They pretty much had enough of all of the mess of a European war, so they decided to have a go by emigrating to a country halfway around the world that they didn't know much about, [where they] didn't know how to speak the language and didn't know what the culture was. And then, their new lives involved cutting sugarcane by hand for a long time for not very much money."

Through his ancestry research, he learned many details about life for his family. He describes his astonishment at their courage: "My grandparents left all of their family behind back then—and never set foot in the country again," he says. "After thinking about it, that's a big deal—packing up everything from what you know, permanently—for the chance of a better life. Being transplanted from one culture to another, without the aid of being able to do any research or google stuff, would have been a major culture shock. And while I really like Italy, I'm glad I live in Australia."

"They saw their lives as too bleak," he says.

But there were some quirky tales, too.

"When my grandfather arrived in Australia, he was issued meal vouchers on a train trip," says Bianco. "When he handed in his voucher for a breakfast, he had no idea what Corn Flakes were because he had never eaten them before. He thought they looked like *patatine* (Italian for potato chips.) So instead of pouring the milk in with the Corn Flakes, he crunched his way through a dry bowl of cornflakes—which didn't taste like chips—and then promptly drank the milk separately."

Like many Australians, Bianco found his background multifaceted, existing at the intersection of Australia and Italy. To him, exploration of his grandparents' and parents' emigration journey is a step toward self-discovery. He wants to understand how their trip impacted the trajectory of all their lives.

"I had always been curious about why someone would

want to leave the country of their birth, never to see it again, to venture to another place that was completely foreign and where they didn't know the language. I do know that Italy was a 'basket case' after World War II, and both sides of my family didn't see a future there for their children. This is a common theme of the Italian diaspora around the world. They felt it was easier to leave than stay," he says. "I found things make a lot of sense when you see the place where your family history happened. Going back to the country of your heritage to discover your roots is a pretty common occurrence for Australians."

He adds, "I think the best way to understand a history lesson is to go to the place where an event happened and see it with your own eyes," he says. "It was good to meet my relatives, and I felt like a bit of a novelty because they were interested in what it was like to grow up on the other side of the world, considering that many of them had never ventured far beyond the village."

Bianco says he's thrilled to hand down to his children a glimpse at what the family is and who they were. He's happy to have played a role in affirming the lives of those who came before.

"It was totally worth it because it gives you an idea of how chance decisions can affect everyone's lives," says Bianco. "I mean, if all of my grandparents hadn't taken a risk to leave, I wouldn't have existed!"

## LEGACY LESSON

"For anyone considering undertaking genealogy travel or finding out more about your ancestors, just do it. Research your ancestry then travel there. You will learn a lot more about the country of your heritage. You'll find out a lot about yourself and others—and that people around the world have the same needs and wants."

### Destination DNA: How genealogy-focused tourism is driving heritage travel

Stories of our ancestors and the faraway lands they traveled from used to exist solely in our imaginations and in the kitchen-table wisdom of our grandparents and parents.

But, now, thanks to the internet, these tales are no longer family lore. We can sit in the comfort of our family rooms and research our roots, the research inspiring us to visit the places where our ancestors once lived.

Armed with our DNA test results and knowledge of where we come from, tourists are picking their travel destinations, driving a heritage travel trend globally. Ancestry travel made Lonely Planet's list of top travel trends in 2019.[20]

Ancestry travel companies are popping up worldwide. These boutique companies, as well as the major players—Ancestry.com, 23andMe.com, MyHeritage.com, AfricanAncestry.com, and more—offer group and individual tours with custom itineraries so ancestry seekers can visit a town, research archives, visit churches, see a family home, or meet with relatives from their DNA test matches.

Airbnb recently teamed up with 23andMe to offer heritage travel where, after receiving DNA results, customers can book lodging in their ancestral homelands through Airbnb. In 2017, Ancestry.com partnered with Go Ahead Tours to promote the DNA test and travel packages.

There's a whopping market out there. According to a press release from Airbnb and 23andMe, "Since 2014, the number of travelers using Airbnb for tracing their roots increased by 500 percent, and 78 percent of these trips are taken in pairs or solo, suggesting that these are introspective journeys or an important moment to share with a significant other."[21]

## Identity-Inspired Wanderlust: Longing to walk the land of ancestors propels genealogy tourism boom

*Most years, the months from St. Patrick's Day in March through October bring a slew of cruise ships docking in the Dublin port and coach buses pulling up along the banks of the River Liffey, unloading thousands of tourists. Armed with folders and digital family trees, they descend on the Irish Family History Centre at EPIC (the Irish Emigration Museum). They are on a mis-*

sion: *to dig deeper into their roots and set foot on the ground where their Irish family lived.*

*"Two of the most common things we are asked are to find the site of the old family house or homestead and to find living relatives—the descendants of those that stayed behind," says Fiona Fitzsimons, who heads the team of genealogists at the Centre. "We have a highly interactive experience here because many people want to do their own research. My friend's husband expressed it best when he said, 'I wouldn't pay someone to play a round of golf for me.' So, people come to the Centre to do their own research. We also have a staff of genealogists at the ready to help when they hit a brick wall in their research."*

*In the last 10 years, genealogy has opened up as a hobby, as more records are available online. Fiona's husband, Brian Donovan, is the man responsible for putting a lot of Irish family history online. "Brian is responsible for digitizing/publishing over 150 million Irish records—he's made a bigger contribution than anyone to opening up research to all."*

*"Online publication has had a huge impact on access, especially for people of Irish ancestry living overseas," she says. "Better access to*

*a wider set of records means it's possible to trace almost every Irish family back to the 1830s, or earlier. Once they've done their research and built their family tree, they want to come face-to-face with their personal heritage."*

*"There's certainly a huge increase in ancestry travel," she says. There are many theories about what is driving the surge, and Fitzsimons thinks one is personal identity: "To have a deeper sense of who they are and where they come from. [...] This includes the adult children of first-generation immigrants, but also the descendants of colonial settlers. One of the things that most surprises me is that overseas, the sense of Irish ethnicity endures. It's passed on for generations, whereas many other ethnicities are lost in the melting pot."*

*Fitzsimons and the Irish Family History Centre frequently work with travel agents, tour operators, and Tourism Ireland to do the background research for Americans, Canadians, Australians, New Zealanders—any people with Irish origins who want to plan genealogy trips to Ireland. The Irish Family History Centre has arranged genealogy trips for hundreds of travelers, including former US president Barack Obama, and his vice-president Joe Biden.*

President Obama, whose ancestors emigrated to the United States in the 1850s, found the family homestead with the help of Fitzsimons. In May of 2011, President Obama and First Lady Michelle Obama visited his ancestral home in Moneygall, Co. Offaly. Fitzsimons coached the present-day owner on how to guide them through the house.

"[Former President Obama is] a lawyer and a career politician—stoic and focused," says Fitzsimons. "But once he was inside [the house], the penny suddenly dropped. He looked around and asked what part of the house, if any, was in the same layout as when his ancestors lived there (between 1800 and 1850). We expected this question based on prior experience, and I'd coached the property owner on how to answer it. The two rooms to the front were all that remained of the original house.

"Standing within those four walls, he was suddenly confronted by his Irish origins. So, you know, even a president isn't immune to the genealogy bug."

In 2016, former Vice President Biden came to Ireland with his family. Fitzsimons was asked to research the family's Irish ancestral roots in advance. Her research on Biden's family of county Louth and county Mayo was used to plan the Vice President's travel itinerary in Ireland.

"I think the 2016 trip was especially emotional for the Vice President and his family," says Fitzsimons, who accompanied them around Ireland for five days as their genealogy guide.

"In 2010, the Vice President's mother passed away. She had always been very proud of her Irish heritage. Then, in 2015, before he could make the trip with his family, Beau Biden, the Vice President's son, fell ill and died."

"When Joe Biden came to Ireland in 2016, I believe his visit was in part a leave-taking (farewell tribute) of his mother, and also of his son. On day five, we reached the Cooley Peninsula in county Louth and stopped where the family house once stood. Close by were the ruins of an old church and graveyard. The Biden family walked alone there and took some silent time to themselves," she says.

"When they were faced with their Irishness, both [Obama and Biden] had different responses, but very visceral all the same," she says. "President Obama joked a lot during his visit. But, standing under his ancestral roof in Moneygall, he was brought to silence."

She adds: "For Vice President Biden, his Irish heritage was part of the fabric of family life. [I believe that] he drew on his sense of Irishness for strength and solace in his personal loss."

Beyond the tours, Fitzsimons can be found in the Irish Family History Centre. She leads the team of genealogists who can provide online and in-person consultation for "brick wall inquiries." The team also takes research commissions for Irish and British (England, Scotland, and Wales) ancestry.

"Research online is the entry point for anyone tracing their roots. Some of the best evidence is buried in archival records, land deeds, wills, estate records, and more," she says.

During the COVID-19 pandemic in spring of 2020, the Centre launched weekly live Q&As on Facebook. The live sessions proved very popular and became a regular feature.

Fitzsimons also teaches a course over two years in Irish Family and Social History at Trinity College, Dublin.

## CHAPTER SEVEN

## THE SECRETS WE KEEP:
### When your search goes unsolved

For some, the search to learn more about their ancestors starts as a hobby, to create a family tree that tracks their lineage from centuries past. And for others, it begins with a holiday stocking-stuffer DNA test with no commitment except to send in the test. That is, of course, until the results come in and you're hooked.

Sometimes the search is sparked by gaps and questionable facts in our stories, inklings that something is not right—a grandfather whose life is shrouded in mystery, an adoptive parent's identity, a long-lost half sister, a potentially famous ancestor, a surname that turns out not to be a real name, speculation about a family legend, or heart-breaking lives that you want to better understand.

These mysteries have thrust those of us who feel obligated and compelled to know into the role of master sleuths in real-life whodunit stories as we delve deep into the search for our near and distant past.

Whatever path leads us into our journey of exploration, the payoff for our detective work is life changing as we travel through time with generations of our family before us.

The search can become exhilarating. You've made a hit, uncovered a clue, and gotten a tantalizing glimpse that there is something lurking there. Suddenly, you're on a mission to uncover the truth. Excitement and anticipation start building.

But what happens when our sleuthing hits a dead end—when it literally becomes a cold case? Maybe you've made a promise to a dying relative to find answers that have been elusive. Or you've pieced together your grandfather's life, but there's a gaping twenty-year hole. What happens when the family member we are searching for has seemingly fallen off the path and left no trail?

It can feel deflating and heartbreaking all at once when the answers aren't forthcoming.

## Austin Calling: Searching for the truth and nuggets of hope behind the secret life of my long-lost grandfather

*My Story*

On the morning my eighty-six-year-old mother was diagnosed with a rare and terminal cancer of the duodenum and given only four months to live, she reached out to me with an urgent favor: "Mary, can you find out what ever happened to my father?"

Holding her hand at her bedside in the GI lab of Loyola University Medical Center just outside Chicago, I made a promise to find the man whose name and whereabouts had been, until this moment, taboo, a man lost in the lives of his family and buried in an abyss of absence, anxiety, and fear.

The clock was ticking . . . fast. A journalist by profession, and one carrying the pain and hurt from my mother's tender heart, I became Nancy Drew, the sleuth of Ancestry.com, FamilySearch.org, 23andMe.com, and Google. I was determined, and I set off on a relentless internet and cross-continental pursuit to unlock the clues behind the mysterious disappearance of my Irish-born grandfather Austin McMahon.

The facts were few: he disappeared in 1929,

*at age twenty-six, four years after he immigrated to Chicago, and three years after he met and married my also recently immigrated Irish grandmother.*

*He left her behind, along with my mother, then two, and her six-month-old sister, and a trail of heartbreaking accounts of lives rife with hardships and crushed souls.*

*My mother lived with a fierce determination to sever memories of his existence and the damage of his defection from all of their lives. She died longing to embrace a connection with him, to finally heal and understand why he left.*

*During the days and weeks at her bedside, she confessed that the yearning had gripped her all her life. Now, it obsessed her. Who was she? Why had her father left her? Did she have any other family in the world she didn't know about? Even though she had been lovingly raised by her mother and her mother's brother, Michael "Mickey" Crowley, and had created a beautiful family of her own, she couldn't let go of the idea that something was missing.*

*Unfortunately, the answers did not come in time. She died not knowing, as I bumped into roadblock after roadblock in search of an elusive fugitive who had left few facts along his trail.*

*My failed promise upped the ante. I had to find out, for her, for her sister who, at ninety-one, still longs for the details of a father who left her more than nine decades ago. And for myself, my siblings, my children, my nephew, and my grandchildren. It's a question that has haunted me since childhood, when I started wondering why his name, his existence was a hole in our family story.*

*During the last five years, I have made it my mission to track down my mother's father—a mission that has taken me four times to cemeteries, parishes, and centuries-old cottages in Ireland, England, and Scotland. I've followed (or tried to discover) the footsteps of my grandfather and his relatives through the towns of Portmagee, Clonakilty, and Mountcollins in County Limerick, Ireland; the annals of British naval ships in WWII; and a garden apartment in Edinburgh. I've uncovered treasures (my family links to comedian Jimmy Fallon, whose great-grandmother Honora O'Connell was the sister of my great-grandmother Ellen "Nellie" O'Connell McMahon). The discovery seemed to come out of nowhere when a member of my Irish writing group happened to have the same family name and sent me a chart done by the clan, which included Austin's lineage, including myself, my children, and the popular NBC comedian.*

*That was pretty cool. The dark secrets—a jail in Hennepin, Minnesota—aren't.*

*This has been a personal odyssey. The bits and pieces I dug up were discoveries made by chance. To this day, I still hunt and peck, and just when I want to give up, a sense and a feeling spurs a series of serendipitous events that are completely mind-blowing. In some instances, the learnings come from cognitive dreams that seem to connect me to my deceased grandfather's soul.*

*I owed my mother the best attempt at getting the truth.*

*I was looking for what happened, but it was the why that matters.*

### My grandmother: Bridget

I knew I wanted to be a writer by age six, and my grandma was my first storytelling coach. Born on a peat farm in Bonniconlon, County Mayo, Ireland, she had limited schooling, but she devoured the *Chicago Tribune* and the *Sun-Times* at the bookends of each day.

She could create whole worlds out of one sentence.

My most vivid memories are of the summer when I was nine and we'd just moved across the Chicago suburbs to Northbrook to be closer to my dad's job at Allstate.

School hadn't started, and my circle of best friends existed solely in longing, suspended until the pending September beginning of fourth grade. I'd race home from swim lessons, or babysitting Katie McGinn and her four brothers, jump off my bike, and make a beeline to the backyard. There I'd find her, stationed on her woven green mesh folding chair, clutching her rosary beads and whispering Hail Marys.

Sometimes she would be still, staring at the weeping willow tree and the beds of roses, peonies, lilies, and lilac bushes that formed a hedge around our backyard. I'm convinced that my parents never even looked inside the house when they made the purchase, instead falling in love with the screened-in porch hugged by a slate patio with birches popping out of it and a bountiful botanic paradise at the foot of a forest preserve, which happened to be connected to the house on the cul-de-sac.

Even today, too many decades later, I can still see her face lit with sunlight, almost angelic looking, her light-blue eyes sparkling. Across four generations, I sometimes startle when those same eyes smile up at me from my three-year-old granddaughter, Keira.

In the moments before my grandma heard the latch on the gate and saw me, I'd stop and just observe her.

I'd imagine she was conjuring up memories of her childhood, the stories I loved to hear over and over about growing up in Ireland with her six older siblings—John,

Mickey, Martin, Mary, Katherine, and Tommy—racing under the clothesline playing tag and laughing, with the wind from what I imagined as a child was the sea spraying their faces. Much to my astonishment, when I eventually found her childhood home, I discovered the water they washed their clothes in wasn't exactly the ocean at the foot of the Cliffs of Moher that I had imagined, but actually a brook that ran through their farm. She told me how they would carry baskets brimming with clothes and forge down a steep hill to scrub them in the water, then march back up and hang them to dry. Never did she describe this hoisting ritual as a chore. She made it sound so adventurous and fun. I pictured an Irish version of the von Trapps.

Dancing around the clothesline like a maypole is the most vivid image I carry of how she used to live. Perhaps even as a child, I sensed that in the telling, she was trying to reclaim something.

Spotting her decked in her wide-brimmed straw hat with her knobby sweater wrapped tightly around her shoulders (even in ninety degrees and August humidity), with wadded hankies stuffed up her sleeve, felt like coming home to a beacon, a lighthouse. I feel her embrace and smile today. "God is good" was her mantra. When I envision a loving God, it is a pint-size elderly woman enfolding me in her embrace, a trail of tissues falling like leaves from her sleeves.

Sometimes, her stories would be punctuated with her favorite song—I still remember the words:

> *In Dublin's fair city*
> *Where the girls are so pretty*
> *I first set my eyes on*
> *sweet Molly Malone*

> *Now her ghost wheels her barrow*
> *Through the streets broad and narrow*
> *Crying, "Cockles and mussels,*
> *Alive, alive-oh"*

But sometimes, especially when she sang the song, I would feel a sadness wash over her, like a cloud suddenly shadowing the sun, and she'd turn her head to the side, trying to hide from me as she'd swipe her hankies to brush away a tear. Was she nostalgic for her parents, Thomas and Mary, or the two siblings, Mary and Martin, she left behind when she crossed the Atlantic to follow some of her siblings? Or was it loss or longing for something darker?

I drilled my mother about what made Nanny so sad. "We don't talk about it" was the short version of the door that slammed the conversation shut whenever I queried.

From my mom's responses, I learned to spot the Irish "look"—that stoic facial expression behind the freckles, the stone silence that lingers beyond the comfortable lag between question and answer. That stare.

But one day, my aunt Dorothy slipped and spilled the little she knew about her father, which she'd been told throughout her life not to talk about. Dorothy is my mom's redheaded younger sister, "the beauty," who was an exquisite portrait artist. Sunday noon supper with Dorothy in the nursing home is my new ritual.

Then, once, when my mom was working and I was babysitting my younger siblings, I sneaked into my parents' bedroom—off-limits to me—and found, aha! my mom's six-inch-thick Bible. There, after leafing through the pages, I found a lone photo, simply marked "Austin McMahon," . . . and staring at me from the page was a male version of my mother, dressed in what looked like a military uniform. The jet-black hair and blue eyes . . . my mom and me. "You and Issy and Austin, you're all clones," Aunt Dorothy always tells me.

I stared at the photo for what seemed a long, long time, then carefully tucked it back into the pages. A couple of years later when I went to peek again, it was gone. I pored through every page, but it had disappeared. I would not see it again until five decades later, after we buried my mom and I was sorting through her belongings. I wonder even now if she knew I found it.

"Of course Nanny was sad. Austin, my father, left us when we were just little girls. Nanny always cried, but she would NEVER talk about Austin. Not bad, not good. NEVER," Dorothy tells me, over and over.

As a child, I had a dream that I was in a beautiful, stone-walled garden and a handsome man, who seemed old to me then, but probably was forty-something, with raven-black hair and blue eyes, approached me smiling and reaching out his hand. I didn't exactly know who he was, but I felt safe. I thought he must be my grandfather. But since I knew nothing about him, I just kind of shrugged it off.

After finding that photo of him in the Bible, I knew Austin was the man in my dreams, the man in the garden.

Despite the icy silence around the Austin issue, I was able to pluck some facts from my relatives:

My grandma, Bridget Cowley McMahon (the name Cowley had been accidentally changed to Crowley at Ellis Island), also changed her name to Virginia to Americanize herself when she arrived in New York in the early 1920s. She met Austin, also newly from Ireland, at a dance in Chicago. They married in 1926 at Holy Name Cathedral and immediately had three little girls. The oldest, Mary, died shortly after her birth. He disappeared when Dorothy, the youngest, was six months. My grandma became a nanny, caring for other people's children to put food on the table for her own.

That's all I knew, even on the day I said goodbye to Nanny, who was living in my parents' room that summer with an oxygen tank, because she wasn't SICK, but just had "a little bug."

On the day I was headed across the country to Arizona State University and freshman year, we hugged and she held on to my hand, almost tugging it, and she said, "Go get 'em. There's a whole world out there waiting for you, Mary."

We landed in Phoenix, and my mother called home. My grandma had died while we were in flight. My sister and mom hopped on the next plane and flew home.

### The Search for Austin Begins

As a child, I recorded nuggets of my grandmother's stories on the pages of my mind. And after she died, I found the photos, the letters from her brother Martin, from the family farm in Ireland, the birthday and holiday cards she'd saved, penned in childish handwriting from Mary Beth, Paulie (my brother, Paul, who would kill me today if I added that "ie" to his name), and my little sister, Sheila.

Dusty and tattered, the collection of letters to and from my grandmother's siblings helped me gain new insights into their relationships. They would be the first clues to the mystery I promised my mother I would unravel during our conversation from her bed in the Loyola's GI lab.

But nowhere in that collection or my mother's belongings was there a clue to Austin, except for that one photo I had uncovered in my mother's Bible. I also had a story from my aunt Dorothy that Austin had written them a letter in the late 1930s that he would be coming home soon after the war, but they'd learned later he was killed in Dunkirk. (My reporting and records later proved this to be untrue.)

Ancestry.com and the genealogy group that meets the last Sunday of the month at the Irish Heritage Center in Chicago would become my new best friends. I swiftly uncovered some facts:

- Austin was born in 1903 in Mountcollins, Limerick.

- In 1911, when he was eight, he lived with his parents, Patrick and Ellen "Nellie" McMahon, and three siblings on Strand Road, Clonakilty, Co. Cork, Ireland. The census information on FamilySearch.org said, "Son RC RW 8 Scholar Single born in Co. Limerick Irish English."

- In 1921, his father, Patrick, a member of the Royal Irish Constabulary, was killed in an ambush in Youghal in east Cork.

- In July 1924, at twenty-one, Austin emigrated to Halifax, Nova Scotia, Canada, with nineteen pounds and seeking "honest work."

Austin's immigration papers on FamilySearch.org said, "He paid for his own fare and could read. He lived in Toronto, Canada with Michael O'Shea and listed his nearest relative as his mother, Ellen McMahon, Portmagee in County Kerry."

- In 1926 he married my grandmother. Mary, my mother, Isabel, and Dorothy were born in the next few years. Mary died shortly after her birth.

- In 1930, a census document read, "He was living from 3/5/30 to 5/17/30 at Third Ave. South & Fifth, Block 36, House #120, Dwelling 32: the county jail in Hennepin, MN. The reason for his detainment: Illegal Immigration."

A jail? The idea that my grandfather was arrested and in jail really stunned me. Despite my searching and writing to the records department at the jail, I still have not been able to discover where he went after. I wonder, did they put illegal immigrants in jail for three months in 1930? Or was there another reason for his arrest, and what was that?

The trail of clues seemed to end there. Despite numerous internet searches, I was not able to find out what happened to Austin after 1930 when he was released from jail.

Until 2018, weeks before I was to leave on a trip with my Irish book club to Ireland and Scotland. A clue from someone who was also looking for Austin popped up on Ancestry.com. By now, I had expanded my search to Ancestry.com and myriad genealogy websites. I found his death certificate:

"First Radio Officer"
42 Male
British
Mount Collins, Limerick birthplace
Place of Death: Cardiff
Port of Registry: London
Ship: Baxtergate
August 23, 1945
Cause of death: Presumed drowning

It also listed his address. On that trip to Scotland, a cherished friend, Mary Anne (one of the more than 20 Mary-Somethings in our book club), accompanied me in a cab through the rain-soaked streets of Edinburgh across town from my hotel to a row of two flats. Upon further research, I discovered that the street he lived on is located in Edinburgh's New Town neighborhood, today one of its most affluent areas.

There I found the downstairs garden apartment where my grandfather, according to records, last lived.

As I stood there, I wondered what I was seeking to discover about his brief existence and his disappearance in our lives.

What and why was I seeking? Was it a place of peace and forgiveness for my grandfather, so that his story could finally be shared with our family? Was it to close the crevasse left by his secret life? Was I hoping to discover he wasn't some deadbeat dad who vanished into thin air? Could there be, was there a reason he had to run? Did I think I could somehow make him know how much sadness his disappearance caused to my grandma, my mother, and my aunt Dorothy, who still struggles with depression every day?

Was he hiding from someone? Was I trying to resolve a curiosity about my grandmother and my mother, a hunch that there was always a deeper sorrow that lay just beneath the surface?

I didn't know. I turned to race back into the dryness of the cab and glanced across the street. There was a stone-walled, iron-gated English garden. I moved toward it, crossed the street, and looked inside, and for a brief second, I was in my eight-year-old dream again. For a brief second, the handsome man with the blue eyes was walking toward me, reaching out his hand and saying, "Mary, you have to tell them why I left."

And then he was gone.

It has been more than two years since that afternoon. I have been on two more trips to Ireland, one in which I found the home in Kerry where my grandfather's mother had moved his siblings in 1921 following the death of her husband, my great-grandfather and Austin's father.

Through a set of serendipitous events—my aunt Dorothy's DNA test revealed a third cousin who connected me with the man who bought the McMahon home in Portmagee, Kerry, from my great-grandmother in 1940—I found myself standing in the front room of that family home, which was also the town drugstore. But though the owner knew all about Austin's mom and his sisters, he sadly reported that Austin was rarely mentioned and never set foot there. The sisters, he says, died young as a result of a typhus epidemic that swept through and killed many in Ireland.

My sleuthing was reaping some rewards. But despite my searching, and searching, and searching, all clues have now led to a virtual brick wall.

I've learned that people who disappear have their reasons why.

But I promised my mother I would find out. I owe it to her, to her mother, and to our family. I will not give up.

### The Descendants: When ancestors elude us, what to do next?

*Almost daily, Jeanette Weiden gets appointment requests from people looking to launch their genealogy search.*

*As the manager of the history and genealogy*

*department at Loutit District Library in Grand Haven, Michigan, Weiden is often the starting point for people on their own ancestry quest, who sometimes know nothing more than the family name and a few scattered details about their recent and long-past ancestry. In other cases, people get their test results back and aren't sure how to interpret them. That leads them to Weiden, who can help people wade through the confusion and help them satisfy their interest in finding out about where they came from.*

*"I think we all have a desire to learn more about ourselves. In researching your family, you can learn more about yourself," says Weiden.*

*The constant barrage of requests speaks volumes about our growing fascination with and yearning for finding detailed information and evidence of where our ancestors came from. She often uses old-fashioned resources like marriage certificates, passenger lists, and newspaper clips to dig into the past.*

*Weiden describes genealogical research as "like doing a jigsaw puzzle, where you hunt and peck and each piece leads to another like a little prize." She's confident that anyone with some time and effort can find information about their ancestors.*

*Except that her own search kept turning into a dead end.*

*"Sometimes you just can't find the answers, even though you know they are all out there,"* says Weiden. *"It's just about finding where they are hiding. You've got to keep going, to stay on the journey. When you get those moments where you find something out, it brings great joy. It's like the fix that puts you back on the high to find records that you can't seem to find."*

For years Weiden, a mother of one in her midforties, has been employing her expertise to learn the identity of her great-grandfather.

Months before she was born in 1974, her maternal grandfather died suddenly of a heart attack, leaving a trail of mystery and questions behind him.

"My grandfather never spoke about who his father was," says Weiden. "My mom's dad had never spoken about his own father, and she never knew her own grandfather."

Through combing old records, she was able to find a name: Willis Delano. It turns out that he was married to her great-grandmother for a few short years in the early 1900s. But that's all she had about the man married to her great-grandmother. During years of research, Weiden was able to find other men in other states with the same name but was unable to connect them.

With the traditional tools of the genealogy trade coming up short, Weiden turned to a more recent trend—DNA testing.

"A lot of times, genealogists are using this test to break through a brick wall," she says.

She had her mother take a DNA test, and the results confirmed what Weiden had suspected—a man she had identified at various times in Detroit and Texas was her great-grandfather. This was done by looking at her mother's DNA matches and finding people who were descended from those other men in other states. There was no denying it. They were the same man, and all of his descendants shared the same DNA.

She found a man whose trail led from Michigan to Texas back to Detroit and whose history included three marriages and nine children.

"Just with genealogy, looking at the records, I never would have been able to connect them," says Weiden. "But with DNA, there's no question—it's the same person."

Her findings also sparked some forgotten memories for her mother, including two men who showed up at her grandfather's funeral claiming to be his half brothers.

### It's Complicated

Weiden's great-grandparents were only married for about four years, beginning in 1902. Shortly after the birth of her grandfather in 1904, her great-grandfather enlisted in the army and disappeared. She's searched for a divorce record but hasn't been able to find one. Her great-grandmother remarried in 1907. She went on to have several more marriages. The great-grandfather, Willis Delano, moved to Texas, got married, and had a daughter. While he was away from home, he heard his wife was "running around town," and he left her and his child again. He showed up a few years later in Detroit, married, and eventually had seven more children with his new wife.

"It answered a lot of questions for my mother about why her father never talked about his dad and she knew nothing about her grandfather," says Weiden. "Unfortunately, I've tried to reach out to his existing relatives (our relatives), but they are not interested in connecting with us."

As she and her mother know firsthand, finding the results can cause further hardship.

"I don't think everyone grasps the reality of what a DNA test will give you," she says. "It can give you great information, and you can find relatives you never knew. But it can also give you a really big surprise that can really devastate your family. There are times when you get a half sibling that you didn't know about, or you find out there's a nonparental event, meaning you find out your dad is not your dad."

Weiden said it's not unusual to find such shocking family secrets. She estimates that one out of every five people she works with ends up finding more than they bargained for. This could also be a cousin or more distant relative, not always just parents.

She once worked with a woman who was confused about her DNA results because they showed that her brother was of a different ethnicity.

"She kept saying, 'My brother came up as a "close relative," but I knew there had to be something wrong with that,'" says Weiden. "So, I asked her to pull up the records on the computer, and unfortunately I had to be the one to break the news to her that her brother was her half brother. Her dad was not her dad. Can you imagine what that must feel like to discover your father is not your father?"

At that moment, Weiden became a counselor.

"There were a lot of tears," she says.

For that reason, Weiden cautions people to think twice before taking such tests. "You never know what you're going to find. There are sometimes unpleasant surprises," she says. "You have to be ready, I think, to deal with those surprises if they come up."

That doesn't mean there aren't positives. She's also worked with people who have been adopted and reunited with their biological parents.

"There's no such thing as a closed adoption anymore," she says. "If you have any fears about what you might find out, I tell people not to do the tests. You have to be ready."

**LEGACY LESSONS**

"I think every person has the right to explore their ancestry and to connect with the people they find, even if it is just one conversation. Whether you are the child or the parent in an adoption, I think you can have a respectful conversation and connection. It doesn't have to lead to a relationship or celebrating Christmas together, but it can bring a lot of peace and understanding on both sides."

---

**Private Eyes: Don't let "brick walls" break you**

In recent years, headlines have pointed to the popularity of law enforcement officials using online DNA results to solve cold cases. Now, family members frustrated that they've hit a wall can take a short break from their research and turn it over into the hands of real-life detectives.

In many cases, like Ancestry.com's "Expert Connect" service, hundreds of genealogists are available for hire to assist with research goals. But also teed up to unlock the shackles of shame and secrets are a growing number of private eyes who are being called on to investigate and uncover the suspects—adoptive parents, MIA relatives, and even hard-to-decipher DNA patterns.

Liz Brock, founder of Root Investigations in Carle Place, New York, is experiencing a surge of calls for genealogy investigations, which include locating a birth relative such as a parent, child, or sibling.

"The shock typically moves to happiness," says Brock. "People realize they just gained an entire whole new family. It can be very rewarding to help people find out who they are and reconnect with their families."

She says 80 percent of her cases end happily. But in a small percentage, "it's like a bomb dropped," she says.

### Serving as a "Buffer" for Bombshell News

Etiquette issues also play a significant role in what she does.

"It's a tricky situation when suddenly an unknown relative pops up, and you're faced with contacting them," she says. "It's kind of hard to say, 'Oh look, I'm your long-lost sister.' An investigator is a little less threatening as a third party. I become a buffer until the person is ready to connect and talk. Most do say yes, they are interested in talking, or they might say they need some time to let it soak in. People need time to digest the information."

She recommends people who find shocking discoveries about relatives they never knew should take this more kind and gentle approach. She suggests reaching out through an email versus starting with a phone call.

"Once a person finally finds their long-lost relative and are ready to make contact, it's a very profound point in both parties' lives, which needs to be handled carefully," she says. "There is a mix of hope and fear bundled into one, and I always tell my clients to expect the worst but hope for the best as we tread lightly into this uncharted territory."

She adds, "Because we, the investigator and client, do not know where the long-lost relative's perspective or mind is at that point in time, we try to approach the situation with kid gloves. We do this first by using discretion and giving the receiver of the information plenty of time to answer questions, sit on the information, and reflect on what they want their next steps to be. Some individuals are extremely excited and jump right into the process, while others need time to think about how they want to handle the scenario."

Brock points out that everyone is different and that by having a third party approach the person in a nonthreatening, factual manner (while providing an ear to listen), you can ease the stress of the situation.

She recommends that the investigator make the initial contact.

"If the results are positive, we can open up the option of the client and the relative making communication," she says. "If it is negative, we act as a buffer for the client to deliver the news. We always deal in facts, but do help make the 'blow' of someone wanting no contact a little bit easier than if the adopted individual contacted the relatives themselves."

Brock strongly advises not to make the initial connection in person.

"Showing up unexpectedly bombards the person with information, backing them into a corner," she says. "You are not giving them time to process the information they just received. I personally avoid this, because I believe this method sets one up for failure."

For many people, including myself, as diligently as we try to research the whereabouts of our ancestors and the past, the information is not easily, or sometimes never, forthcoming.

Hiring a private investigator ranges in price, beginning at about $250, depending on how much information you have on the person you are looking to find, says Brock. But it can cost up to several thousand dollars, she says.

- A private investigator for genealogy is hired to do the following:

- Recommend the steps you need to take in your search for a missing person/family.

- Review your family tree and recommend next steps, i.e., DNA testing. However, she cautions that private investigators rely heavily on records.

- Analyze the information that is found in the records.

- Describe the findings in terms that make sense for you.

- Create an actionable plan for reaching out to the long-lost relative.

- Research ancestral family trees and potential medical issues (for example, families who have a history of people dying from heart conditions).

"A private investigator is different from a genealogist because they have the ability to access particular databases that are not available to the general public and genealogists," says Brock. "Genealogists are solely relying on public records."

Not all private investigators do genealogy research, she says, and she recommends you do your homework when seeking to hire one. She also recommends hiring a private investigator from the state and the country where you think the person lived.

### Lost and Found: Missing years of family history reveal surprising results

*Hazel Thornton likens genealogy research to working a jigsaw puzzle. It sounds fun at first when you start sifting through dozens of pieces. Except, imagine many of the puzzle pieces are missing or broken and there are extra pieces thrown in from someone else's box, and you don't have any of the boxes with pictures on them to guide you. It can be challenging but rewarding when the pieces start to fit and a picture starts to take shape.*

*That's exactly what happened to the Albuquerque, New Mexico woman who combined her background in engineering and fine arts with her love for genealogy and organization into a full-time vocation helping other piece together their family stories through her business, Organized for Life (www.org4life.com).*

*When she encountered a roadblock in her own online research, she hit the books and uncovered a family-tree saga that had been lost for decades.*

*The results were surprising. Her first discoveries came from tracing the family of her father, a retired Veterans Administration Christian chaplain, to the eighteenth century and the era of American explorer and frontiersman Daniel Boone. Her ancestors, like Boone, belonged to the Quaker church, the Society of Friends, and like Boone, they married "worldlings," or non-Quakers, and were kicked out of the Quakers. The Quakers were a group who immigrated to America when their members faced religious persecution in England for their beliefs.*

Like many families, the Thorntons had some significant gaps in their family history. She threw the little knowledge she had about her family history into the puzzle mix and let her consuming curiosity drive her efforts to discover more.

"My dad grew up knowing little about his family history except that they were reportedly of Scotch-Irish descent," she says. "His father had nine siblings, but the only one my dad really knew was his uncle Melvin. Why? Was there a falling-out? We may never know."

But thanks to a little packet of family history that this lone uncle left behind, Thornton learned the names of four of her sixteen great-great-grandparents.

Through her research, she discovered the Thorntons were Quaker pioneers. Everywhere they went, from North Carolina, to Ohio, Indiana, Iowa, and California, she uncovered history books depicting the Quakers as settlers, and within those narratives she found tiny glimpses of the Thorntons' role in that movement.

"It was news to me and my dad, too, how a retired Christian minister and VA chaplain had so much in common (their spirituality) with his ancestors and didn't know it," she says. "How do things like this happen? How does history get lost?"

Just two generations separated her father and his Quaker great-grandfather, Calvin Thornton. Like Daniel Boone's family, most of Calvin Thornton's generation fell away from the church and were disowned for marrying outside their faith and for similar infractions.

"Quaker disowning is not as harsh as Amish shunning, in that one is not ostracized by one's family," she says she learned. "Still, at least in my dad's lifetime, in his branch of the family, Quakerism was apparently never mentioned again."

### Scouring Library Shelves for Clues

Like many of Quaker ancestry—whose birth, marriage, and death records were kept internally and not at county courthouses—she found the Thornton history buried in a history book. She discovered her useful information in a Polk County, Iowa, history book.

There she found a biography of Calvin Thornton, with one throwaway line that zeroed in on the lost family secret: "He was reared as a Quaker."

"One thing led to another—each step of the journey was a story in itself—and I discovered Calvin's parents, Nathan Thornton and Charity Cook, who married in 1821 in Indiana."

Even though Charity Cook Thornton would have known her namesake grandmother, Charity Wright Cook (who died the next year) and would have been quite aware that she was a well-known traveling Quaker minister, the information dissolved into an abyss. Also interesting is that Charity Wright Cook juggled her ministry duties while raising eleven children, says Thornton.

She also uncovered a book, *Charity Cook: A Liberated Woman*, by Algie I. Newlin.

"I wonder if Charity and Nathan knew that their Quaker grandfathers had fought (despite the Quaker pacifist tradition) on opposite sides of the Revolutionary War," she says. "Did they know and not care? Was it still talked about, or was it water under the bridge? Had that episode of their family history already been lost?"

Her father was very interested in the Thornton branch of the family. But her research didn't stop there, and Hazel explored the Clay, Harris, Wilkins, Byrd, Pearson, Taylor, and Henderson families, too, but to no avail.

Until she convinced her father to take an at-home DNA test.

Her father turned out to be 28 percent Irish/Scottish. The DNA test revealed the Thorntons are 98 percent from the British Isles (Ireland, Scotland, England, Wales, and northwestern Europe), and 5 percent Scandinavian.

Unfortunately, her search to identify which ports her ancestors sailed from in their immigration to America still continues.

"Sometimes, it seems I will never identify the immigrant Thornton who first came to America," she says. "As far back as the mid-1600s, all of my Thornton ancestors were born in America."

She wishes she could trace even further.

"It's odd that the unproven Scotch-Irish family legend persists, whereas the now thoroughly documented Quaker connection (pieced together from Quaker records) was almost lost." Had Hazel Thornton not researched Calvin Thornton, the fascinating line of Quaker settlers would have been a forgotten piece of family lore.

## LEGACY LESSON

Thornton says she is determined to organize all the puzzle pieces of her family legacy so the history will not be lost again.

## CHAPTER EIGHT

## BE LONGING:
### Bittersweet reunions bring joy, and tears

Feelings over the popularity of at-home DNA tests can vary from folks who say they will never put their personal genetic information out into the cloud to others who claim it's been a revolutionary experience finding long-lost ancestors and better understanding their heritage and health risks.

The one thing that's certain is that there are no more secrets. That's pretty destabilizing for those who are discovering earth-shattering news.

"Even people who say, 'I don't want to know,' secretly deep down inside really do want to know," says David

Rencher, chief genealogical officer at FamilySearch International and director of the Family History Library in Salt Lake City (read his story later in this chapter). "This can be especially true for people who are adopted. But if they wait too long, their parents can be deceased and the closure they are seeking can't be there anymore."

The reality, Rencher says, is that DNA tests have "removed the veil of secrecy."

"If you have a secret you were hoping to hide, there are no secrets anymore," says Rencher. "This is changing the fabric of the human family, because the DNA does not lie. As all of the DNA testing companies say, 'You have to be prepared for unexpected outcomes.'"

In this chapter, Rencher and others who have discovered the unexpected—either through DNA tests or traditional genealogy research—share heartwarming (and sometimes unsettling) stories of adoptees meeting their birth families or people discovering a sibling they never knew existed.

The families who generously shared their stories in this chapter offer an honest, relatable, and heartrending portrait of how ancestry searches and DNA discoveries are affecting families across the globe.

**Out of the Past: Professional genealogist struggles to crack his toughest case—his own**

*David Rencher is a man who makes his living scouring documents to dig up the details of other people's pasts and interpret their elaborate family trees.*

*As the chief genealogical officer at FamilySearch International, director of the Family History Library in Salt Lake City, and vice president of development for the Federation of Genealogical Societies, he knows his stuff. His efforts were lauded at the beginning of 2020 by the American Society of Genealogists.*

*Since he was a college student at Brigham Young University and switched his major from business to genealogy, the family story detective has been passionate about helping people discover their ancestors.*

*His journey has taken him online and offline to libraries, churches, courthouses, and other archives across the world, where he's searched the paper trails of others' ancestors through meticulous examination and analysis of documents, artifacts, and other clues in order to help people locate biological family members.*

*Sometimes, genealogy detective work can mirror the popular TV shows and Netflix dramas where cases go unsolved and elite special squads are discharged to investigate these unresolved mysteries.*

*These are called cold cases, and Rencher's family story is one of them.*

This is his story.

In 1953, David Rencher was born in Pennsylvania, and the family moved "very quickly" to Arizona before he was one. He, his parents, and his two half siblings—his sister, Connie, then thirteen, and eleven-year-old brother Greg—moved to Phoenix. When he was eight years old, his mother died of ovarian cancer. At some point during his younger years, his father had left the family and his mother was raising the three children on her own.

At the time of his mother's death, Rencher's former first-grade teacher intervened and suggested to his father that it might be best for David to be adopted. His older sister was already married and had her own family, and his brother was graduating from high school. His brother would later join the military.

The process of adopting the second grader took a couple years to finalize, as the adoption agency was trying to track down his father to get permission. David was ten when he was adopted by his teacher's sister and her husband, and he moved to join the family in Winslow, Arizona, where he started third grade.

"That changed everything in my life," says Rencher. "I was suddenly going to church" (his adoptive parents were Mormon). His adopted father worked for the Santa Fe Railroad, and the young Rencher grew up as the only child in the household, playing trumpet, attending high school, going to college at Brigham Young University. He got married and then joined the Navy Reserve.

On the morning after his wedding in Washington State, his sister Connie said to him, "We need to talk. We have another sibling from our mother who was adopted at birth and is a couple years older than you."

## The Search Begins

That statement, "we need to talk," would start a more than twenty-year journey that would lead to the discovery of several surprises and a half dozen siblings.

At the time, Connie also handed Rencher a photo of a girl about ten, who was holding an infant on her lap, and said the older girl was Linda, a half sibling on David's dad's side (Connie and David did not have the same father). Connie had obtained the photo when their mother died and she inherited their mom's photographs. At one point in her childhood, Connie had met the older girl in the photo.

Putting his professional genealogy cap on, Rencher started the search for the half sibling on his mom's side. His only clues were that the father was affiliated with a particular baseball team, that the child was a couple years older than him, and that there was either a Pennsylvania or a New Jersey connection.

"I poked around and found nothing," says the father of five and grandfather of four.

Another twenty years went by, and Connie stopped at the Rencher home, now in Utah, on her way to New Mexico, where she randomly mentioned that the child was adopted through Catholic Charities in Philadelphia.

That's when the case cracked open. With that information, Rencher was able to discover through Catholic Charities that there was indeed a sister. They told him they would deliver a letter to her and if she responded, he could connect with her. The sister responded and told him she had been adopted.

"I learned then that someone could be sitting on a clue that could break the whole search open, if you only knew the clue," says Rencher. "To think that if Connie had said what she knew twenty years before . . ."

Now determined to uncover a potential half sibling on his father's side, the girl in the photo, Rencher traveled to Pennsylvania to investigate whatever records he could find. His search didn't bring up anything, but having worked in a funeral home as a teen, he remembered that funeral homes could be a good source of information. He started a tireless phone search and found the death notice of his father's first wife, through which he learned of a daughter, Elaine, who now lived in Pottsville, Illinois.

"I picked up the phone and called Elaine and started by saying, 'I hope this is not a bad thing, but I think I'm your brother,'" he says.

He was surprised by her response. "Oh my God. Your birthday is in April, isn't it?" Rencher responded "Yes." Then she said, "Mine is in June of the same year." Rencher says he suddenly understood why the family moved across the country so quickly when he was a baby. It turned out that Elaine had been searching for David during the same twenty years he'd been looking for her.

Elaine was the baby on the girl's lap in the photo, and she told him about Linda, her older sister.

"My father left his first wife with the two girls for my mom," he says. He explains that the reunion was a little harder for Linda, because he was now the age his father had been when he'd left her, and Rencher looks very much like his father.

"Finally, I go see Linda, and she has a beautiful family, and we sit around the dining room table having a nice chat, and she shows me photos of our grandparents on our father's side, who I had never met before. She and her sister had been in constant contact with them even after my dad left them."

"We're sitting around the table, and they are like playing Ping-Pong with their eyes, looking back and forth at each other, and finally my brother-in-law says to her, 'Are you going to tell him?' She then tells me we have another half sibling on our father's side. I felt like this was déjà vu."

Rencher, who spent his junior high and high school years growing up in an only child household with his adoptive parents, now realized that he has six half siblings (three on his mom's side and three on his dad's).

Though he has launched a search through online and off-line resources, he still has not been able to discover the whereabouts of the final half sibling. But he vows not to stop searching.

Throughout his journey, he says he has been very blessed by the discoveries and other surprises that he has encountered along the way.

In the late 1990s, when he was sitting at his desk at the Family History Library, a man popped his head in the door. The man said he saw the sign on the door with his name and ventured in to ask, "Are you from the Rencher family in Arizona?" The man, Melvyn Thomas *Shelley*, a former Arizona Appellate Court judge in Arizona, was the judge who had handled Rencher's adoption.

"It was complete serendipity," says Rencher. "Judge Shelley told me he was visiting the guy in the next-door office and just happened to see the sign on my door. He said he was so thrilled to see that I had turned out so well, that it was a lovely moment for him."

## LEGACY LESSON

"Everyone's story is different, but it is in the gaping holes where, over time, families are reunited and they discover their stories. You just must be very, very patient and not give up on those gaping holes. Here I was with all this expertise in genealogy, and still I couldn't get there to the answers. The most insignificant facts can be the most significant."

### She Longed to Meet Her Biological Parents: How a DNA test leads adopted woman to Korean parents

*Mallory Guy knew from a very early age that she was adopted. Her mother and father provided an incredibly loving home in Parma, Ohio, where she and her four siblings —three of them also adopted—thrived.*

*Guy never felt out of place because her parents, Jim and Pam Knauss, took great care of her and her two older brothers, Jeff and David, and two younger sisters, Ashley and Amber.*

*Tracking her original parents proved to be difficult, as a page from her adoption agency file was missing. Her adoptive parents were told that as an infant, Guy was left outside the orphanage without any useful information. Her Korean name was given to her at the orphanage, and even the date of her birth was estimated.*

But the little girl who loved to read always imagined what it would be like to meet her Korean parents. She had a deep longing to find out more about who they were and what their life was like in the country where she was born.

In 2013, Guy, thirty-two, who was now a mom to Emmie, four, and Jordan, two, decided to take a DNA test to see if she had any genetic health red flags.

"I really had no expectations about trying to find relatives, because I thought I had been abandoned, and so what was the point," the Mentor, Ohio, resident said.

It's not uncommon for adopted children who seek their biological families to run into many roadblocks and untraceable birth records. There's also the possibility that the biological parents who made that decision long ago do not want to be identified, much less contacted.

Fast-forward, and after six years of frequent monitoring and frustration over lack of any matches, a close DNA match showed up in Washington State.

The woman, April Johns, turned out to be Guy's second cousin, and her mother was Guy's great-aunt.

On September 3, 2019, the duo connected her

*to her biological mother and father and her two*
*siblings in Korea. "I was shocked," she says. "I*
*was at work when I got the call and just couldn't*
*believe what I was hearing."*

After a lifetime of longing, Guy was connected to her biological father, Ki Yeol, and mother, Mi Soon, with her
great-aunt's help. She also learned about her older brother
Injae and her younger sister Inseo.

And when she finally made that connection, the myth
that she had believed, that she was simply abandoned, was
replaced with a new, life-giving truth. Guy was born with
a cleft palate, and Guy's biological parents had made the
tough decision to give her up for adoption to Americans
who could afford the extensive surgeries she needed.

"I had believed all my life that I was abandoned at an
orphanage at four months, but now I know that my parents did it out of love to give me a better life," says Guy.
She learned her Korean birth name was Jae Boon Lee.

Her adoptive parents were very supportive of her discovery, she says.

Guy's adoptive parents were German and Italian,
and the area of town they lived in was predominantly
Polish, so the delicacy was pierogi. Now she's learned
that her Korean biological parents own and operate a
small restaurant, and that every year on her birthday
in March, they celebrate it and think of her by making

seaweed soup, a Korean birthday tradition in honor of loved ones.

In March of 2020, Guy and her American family had planned to travel from Ohio to Korea to meet her birth parents and biological brother and sister. During their three-week stay, her Korean parents had planned to cook for her, and she'd already scoped out a Korean grocery store in town. But then the COVID-19 pandemic hit, and Guy says the family has postponed the trip until 2021, or whenever it is safe to travel again.

Since their reunion, Guy speaks to her parents (through a translator app) at the beginning and close of each day. "It's like a fairy tale, and probably the best way an adoption story can go," says Guy. "In many ways, it is overwhelming because I have such a loving family and mom and dad here, and now I have another family in Korea who want to be part of my life. It's like we are catching up for lost time. I feel very blessed."

## LEGACY LESSON

"I think the most important lesson I learned is not to make assumptions about what people think about you. I assumed my parents just didn't want me and abandoned me, but it was way more complicated than that, and not true. We never know what someone's whole story is and what they are having to deal with. It's been a life-changing lesson and experience."

## Guess Who's Coming to Dinner? Finding new family members through DNA test

*Home ancestry tests may tell us more than we wanted to know. That was the case for Kat and her extended family.*

*When her aunt and uncle joined the 28.5 million people worldwide who had submitted their DNA for testing, they weren't expecting to learn much. They took the DNA test in search of more information about their heritage. But what they discovered was two female relatives they didn't know existed.*

*After further investigation, they discovered the young women were their son Kevin's daughters, granddaughters they never knew about. Neither of the girls knew about the other, and their mothers never told Kevin they'd had his child.*

*Stunned, and certain the test had made a mistake, they went from shock to nervous excitement.*

*For many families the fact that there are surprise siblings, half siblings, or offspring can be blindsiding. It also opens myriad questions about the right to reach out to these unsuspecting relatives. Sometimes these attempts are met by anger or silence.*

*But what happened next in Kat's family is*

*testimony to the family's determination to welcome and embrace all their living descendants.*

Every year since she can remember, Kat and her large extended family—more than 150 great-aunts, uncles, and cousins—hold a family reunion in Atlanta. These homecomings are all about getting together and celebrating the family with an impressive lineup of games, foodstuffs, and traditions.

Despite the large ranks of relatives, "Everyone knows the family tree like the back of our hand," says Kat. Her grandmother was one of twelve children, so it has been her lifelong mission to keep the family and its future generations connected.

Shortly before the annual event, her aunt and uncle, both in their seventies, decided to take a DNA test to learn more about the family and its history.

They had no idea what was in store.

Their discovery was not about their parental lineage—they discovered they had two granddaughters they never knew about. It was a secret that their son, Kevin, fifty-one, didn't know about, either.

The ancestry test revealed the two younger female relatives. The thirty-something women were Kevin's biological daughters, neither of whom had ever met Kevin, since he didn't know they'd been born.

The strength of Kat's aunt and uncle, and her cousin

Kevin's commitment to welcome the daughters into the family, came immediately to light. They were immediately welcomed into and adopted by four generations of family gathered at the reunion.

"I must admit it was kind of shocking when my cousin, who I've known all my life since we were very young, came up to me and said, 'Hey, have you met my daughter,' about the young women standing next to him," says Kat.

## LEGACY LESSON

Since her cousin and the family made the discovery, "It's been a lovefest for our whole family," says Kat.

### It's Never Too Late: Sisters find each other and the friend they never knew they had

*It's never too late to experience a great many things—and this includes reuniting with a long-lost sister. When Sherry Gavanditti's cousin said she was researching their family ancestry, the Bedford, Ohio, mother of two grown children really didn't think much would come of it. In fact, she says she really didn't give it much thought afterward.*

*"I [had] a rather tumultuous childhood, and so the idea of looking back at family didn't really interest me, because I thought I knew all there was to know about my mom and dad's side of*

*the family," says Gavanditti. "I was happy. I had a husband of thirty years, two grown daughters, and a nine-year-old grandson. I'm super busy with my work and a bunch of projects I am working on, and really just didn't have time to explore my ancestry."*

*Gavanditti had no idea that her relative's simple DNA test would uncover that she had a forty-seven-year-old "baby" sister she didn't know existed.*

*"I was shocked, utterly shocked," says Gavanditti. "I remember I was at work and my cousin messaged me to say that she found another first cousin, and that it looked like that first cousin was my sister. It took me a while to have it really sink in. It was surreal."*

*Within twenty-four hours, Gavanditti and her cousin had tracked down the sister and reached out to her on Facebook. Shortly after, her sister messaged her back. Her sister was equally shell-shocked. Initially, her sister's response felt curt: "This is hard to handle. I don't know what to say. I need some time to digest this. I'm sorry," her sister wrote back.*

Fast-forward seven months, and the sisters have discovered a whole new world of what it means to be family, says

Gavanditti. It turns out that after Gavanditti's father and mother split up, he had a daughter—ten years later. But in a freak turn of events, he died before she was born. No one in his family knew about his unborn child.

The daughter, Stephanie, was raised by her grandmother thirty miles away. Once the two sisters reunited in August of 2019, the coincidences that seemingly tied their sisterhood bond together began spilling out. For instance, both sisters have husbands named Kenny. Both were born in Virginia.

Today, Gavanditti is fifty-seven and Stephanie is forty-seven. Gavanditti lives in Ohio, and her sister lives in Kentucky, just across the Virginia state line.

Once Stephanie's initial shock wore off and the long-lost siblings agreed to meet, Gavanditti made a beeline to Kentucky, driving eight and a half hours to meet her little sister.

"I remember asking my husband if we had a fishing pole anywhere and he was like, 'Why do you need a fishing pole?'" says Gavanditti. "I'd been texting my sister and learned she loved to fish and that she was planning on taking her stepchildren fishing the next morning." Stephanie didn't know her sister planned on making the road trip until that evening. "I texted her asking, 'What are you doing tomorrow?' She said she was going fishing. I thought it would be great to surprise her, so, I wanted to bring a fishing pole to fish with her."

The duo hit it off immediately and made some discoveries about each other. While Stephanie was an only child, Gavanditti was raised with two younger half brothers.

"You have my chin," Gavanditti remembers both of them almost simultaneously blurting out. "We didn't go fishing; instead, the kids went with their dad, and Stephanie and I sat at the hotel on the back patio, shoes off, relaxing as if we'd known each other forever. Suddenly the same thing came out of our mouths: 'You have my toes.' We cracked up. It was so weird! Having someone look like you!"

That first night, Gavanditti says they spent hours and hours talking, trying to catch up on their entire lives.

"It turns out we have so, so much in common," says Gavanditti. "She says that she always thought she was so alone but that now she's so joyful to have discovered that we have each other. For me, I almost feel maternal toward her, like she's my little sister and I want to take care of her and make up for all the years I couldn't."

"She's just the cutest little thing," says Gavanditti. They're planning another reunion in the summer of 2020. This time Stephanie and her husband are going to visit Gavanditti and her family in Ohio.

### LEGACY LESSON

"I think the biggest lesson I learned through all of this is that we don't know what we don't know. But we have to be open and willing to accept surprises that come into our lives. When we do, there is an opportunity to welcome wonderful new people into our lives. I also realize that not every family is going to have a happy ending and that you need to be cautious when you are reaching out to contact a long-lost or never-known relative."

## It's Complicated: Adopted at birth, two sisters find eight siblings

*Born exactly a year apart to the same biological parents and adopted separately and in secrecy at birth, Wynn and her sister Erica found each other in college through adoption records. They bonded immediately, recognizing the same deep, sultry voice they both shared when they intro- duced themselves on the phone.*

*Wynn, forty-nine, who grew up with an adopted brother near the New Jersey shore and was raised by a very loving single mom, always wondered whether she had any other biological siblings. When she discovered Erica, who moved from New Jersey to the South at nine, they shared rumblings of information they'd gathered about their biological parents, and about maybe having other siblings, but they didn't know for sure.*

*That is, until they connected with their parents and learned they had three older siblings that the parents kept and raised. Years later, DNA tests and Facebook investigating led Wynn and Erica to discover they have more brothers and sisters—seven of them, actually, plus a half sister, adding up to a whopping total of ten siblings.*

*"It's been overwhelming, and many of us are still healing from these discoveries," Wynn says, recalling the many times new siblings have emailed or called her to say they were also birthed by the same set of New Jersey parents.*

As leaders of the pack, Wynn, who lives in Washington, DC, and Erica, who lives in New Jersey, have continued their journey of meeting and connecting with their siblings, nieces, nephews, cousins, and biological parents.

It has been a roller-coaster ride filled with emotional highs and lows, as the sisters have struggled to put the many pieces of their family story together, with siblings who are scattered in states across the country, from the Northeast to Texas and Florida.

Their parents came from "the opposite sides of the track," Wynn says. Their father's family was more afflu-ent, and he was in college on a track scholarship when the first siblings were birthed. Their mom was sixteen when they started having children. With two boys and a girl, the couple got married. A year later, pregnant with Erica, the parents put her up for adoption, and almost a year to the date later, Wynn was also put up for adoption.

Immediately Wynn was welcomed into her adopted parents' family (her adoptive parents were married until she was four). Her adoptive family staged a "Selected Not Expected" gathering and baby shower, and rushed her directly from the hospital to meet the whole clan.

"They were all waiting for me, and I was very loved growing up, but I knew I was adopted and always had that curiosity," she says. In New Jersey, the records are closed, but Wynn was able to do some digging and in 1988 found out the bare facts about her parents.

Further digging uncovered her three older siblings and Erica. (Later, DNA would reveal that there was a fourth half sister, born a year after her oldest brother.)

Wynn and Erica joined forces in 1989 and reached out to their biological parents. Over the course of the next thirty years, they've met with both parents, been embraced by both parents, and been pushed away by both some siblings and the parents. The push-pull has taken its emotional toll, says Wynn.

The 23andMe.com DNA surprises also turned up a new, younger crew of siblings: a set of twins, a boy and girl, who were adopted and raised in Texas; and another younger sister, who was adopted and raised less than a mile from Wynn's childhood home.

"Most of them were so receptive to knowing [about us] and forming a relationship with us," Wynn says of her siblings. "We've all had an instant connection, and it was like we've known each other all our lives."

Pictures of the siblings reveal a striking resemblance. All share almond-shaped eyes and similar distinctive, and very attractive, features. Erica and Wynn also gave birth to their children within weeks of each other.

"It's like looking in a mirror," Wynn says.

Wynn and several of her sisters have grown and continue to grow their relationship. They call themselves "The Sibs Nation."

"In so many ways, this has been a blessing and a beautiful experience that came out of a lot of darkness," says Wynn. "But there is a loss when you are adopted, because you have imprinted your mother's voice, emotions, and being while she was pregnant, and that is your first loss. But for us, to have reached out, connected, and rebounded with our parents, only to keep having it ripped away, is a second awful loss."

## LEGACY LESSON

"Before you start searching your DNA, make sure you are strong about your feelings for yourself and you understand who you are. You need that foundation no matter which way it goes. Don't take it personally if your family does not want to connect. They may be missing all of this, but you can't control their reaction. We're all on our own journey and allowed to have the feelings we have. Seek therapy if you are having challenges dealing with it. It's okay to get help with your healing."

## CHAPTER NINE

**THIS IS US:**

Our role as light keepers
of our family stories

*"The castle's keep is the most fortified part of the castle. The light is our family, our ancestors, our pictures, our stories and our memories. The light keeper is the one, to me, who is charged to share, invigorate, and fortify their family with the strength found in her family."*

—LIBBY LEWIS,
SPECIAL EVENTS COORDINATOR, ROOTSTECH 2019

In the end, our ancestry research is a gift to ourselves, our families, and generations to come. The stories make us uniquely us. Bad things may have happened in our family's past—things that may still affect our lives today. But good things have happened, too. In the face of difficulties, we learn how our strong, awesome relatives faced adversity and pushed forward.

We honor our ancestors' stories when they teach us about healing, loving, and living.

When we remember and share their stories, we can look at those painful times and we can start to realize how ordinary people navigate extraordinary hardship. We can celebrate their resilience, strength, courage, and grit so that all these qualities can inspire generations to come.

To make sense of those stories and to share them is an act of generosity, says Diahan Southard, of Your DNA Guide.

"For me, that is what family history is all about. It's about people that you've never met who are real in your life because you've taken the time to get to know them, because you've researched them, because you've cared about them," says Southard. (Southard shares her experience tracking her mother's biological family in this chapter.)

"You have your very own unique record, and no one else in the world can tell your story the way you can," says Southard. "And that's what makes genetic genealogy so exciting, because we can each participate in our own way."

But what are family members to do with the "shocking" stories they uncover? The silence is over, the secrets have been pulled out of dusty closets and into the light.

Southard suggests we look at our ancestors' lives through a new lens, as survivors, and as people who thrived.

When we transition from focusing on how secrets have shattered our impressions of who we thought our relatives were, and choose instead not to judge them, we can look at who they really were and be inspired by their creativity and resourcefulness in moving through difficult times.

"I think many people initially find these shocking revelations as a form of betrayal because it shakes our concept of who someone was," says Southard. "We need to shift our mind-sets and instead respect and remember that they were people doing the best they could in difficult situations. Our own moral compass might make us think that those were mistakes. But doesn't everybody make mistakes? If we instead look at them through compassion, we can love them for who they truly are, not for someone we have made up in our minds that they should be."

### My Mother, Myself: A work project turns into joyful family reunion for her mother

*Diahan Southard was hit by the DNA bug in high school and went on to study at Brigham Young University, where she earned a bachelor's degree in microbiology and worked at the Sorenson Molecular Genealogy Foundation.*

*In 2001, when Southard was a student, she*

*was working on a project for the foundation to create the first genetic genealogy database, to try to show that, using DNA, we could tell anyone where they had come from. That's when her mother raised her hand to volunteer to take a DNA test. Interestingly, her mother was adopted.*

*Her DNA journey officially began. Now she had the results in her hands with a long list of women in their lineage.*

*She and her mother really weren't all that interested in trying to uncover her mom's biological parents. Her mother had felt lucky enough to be adopted into a great family with parents who loved her. She was concerned about how it might make her adoptive mother feel, so for a long time, they didn't search for her biological roots.*

Fast-forward through years of improved DNA testing methods: DNA test results from 23andMe.com brought them a list of genetic matches in North Dakota and then Washington, where her mother was born and grew up. Diahan reached out to one of the relatives, and by working together, they were able to identify one set of her mother's great-grandparents.

From there, they researched her great-grandparents and learned that they were part of a large family group of German-speaking Russians who immigrated to the United

States and settled in North Dakota. The couple made the journey with eight children under the age of twelve.

The next clue was her mother's birth certificate, which arrived on the Saturday before Mother's Day in 2015. From the information listed on the birth certificate, they searched listed addresses and followed up on any other information they could find.

Six months later, a new DNA match popped up. It was a half brother. The rest is a happy ending, as she and her mother have met and been welcomed by their biological family on her mother's maternal side, and they have pieced together the circumstances about the birth and adoption.

Southard says she doesn't want to delve into too much detail because she is writing her own book about her mother's DNA discoveries. "We had an amazing experience," she says.

Her mother, seventy-one, found three half siblings who were all older.

Diahan's biological grandmother, they learned, was in the throes of extreme hardship at the time, trying to escape her marriage and keep her older three children safe. The family was devoutly Catholic, which made the situation even more complicated.

"Instead of finding this to be a shocking and upsetting situation, I think my mom and her siblings have a greater respect for their mother and her incredible courage and faith," says Southard.

"At a time when it may have been easier to not carry the baby to term, she did. When it may have been tempting to turn against God and blame him for all that was going wrong in her life, she didn't. She stepped up and showed up as a mother and as a woman of faith."

Since her mother was reunited with her siblings, they've all been communicating.

"They send each other Christmas ornaments and keep each other up-to-date on their lives," she says. "It reminds me that it is never too late to begin a relationship. That family ties can even be hastily woven and then prove to be just as strong as those that have taken years to form."

## LEGACY LESSON

"I know we were lucky to have found a family that is so open and welcoming. But I know this journey would have been worth it even if it hadn't turned out as well. Working on this together with my mom has allowed us to have conversations and experiences that we wouldn't have otherwise. So now our bond is stronger. And that means everything."

### It's a Family Affair: Why telling our stories is so important

*Stories open windows to your past, and family stories of specific people and bygone eras offer histories rarely seen in textbooks. They help us to understand the present and guide the future.*

*During the last few years, I've often turned to thoughts of my grandmother Bridget McMahon, an Irish immigrant (see Chapter Seven) who found herself newly transplanted and suddenly left alone raising three young daughters in Chicago.*

*If she were still alive today, I know that she'd have much to tell me about resilience, courage, and pushing forward against all odds. Instead, I draw on the strength I saw with my own eyes and the stories I've heard about her supporting herself and her children as a nanny, having to lug logs up three flights of steps to heat the small three-story West Side apartment.*

*Always, these memories push me forward, make me feel better, and strengthen my resolve to stay on course and stay steady in the storm on days when I find myself faced with the challenges of raising three children solo.*

*Experts call this "the ancestor effect." When we think about our ancestors, it can have a positive impact on performance on intelligence tests and make us feel better, according to a 2010 study at the Universities of Graz, Berlin, and Munich.*[22]

Peter Fischer and his colleagues have shown that thinking about our ancestors reminds us that, as humans who are genetically much like our ancestors, we can successfully overcome incredible challenges and hardships, just as they did.[23]

The researchers asked eighty undergrads to spend five minutes thinking either about their great-grandparents or about a shopping trip they had taken recently. The results: the students who thought about their ancestors said they felt more confident about their performance in upcoming exams than the students who thought about shopping. Researchers said that is most likely because thinking about their ancestors' lives made them feel more in control of their own lives.

As reported by the British Psychological Society's *Research Digest*: "'Normally, our ancestors managed to overcome a multitude of personal and society problems, such as severe illnesses, wars, loss of loved ones or severe economic declines,' the researchers said. 'So, when we think about them, we are reminded that humans who are genetically similar to us can successfully overcome a multitude of problems and adversities.'"[24]

---

**Senior Moments: Stories bring meaning to the lives of seniors**

Research and anecdotal experiences tell us that learning more about the stories of the elderly in our lives is good for them, and good for us, too.

Along with researchers, organizations like StoryCorps have launched intergenerational programs like the Great Thanksgiving Listen to inspire young people to capture the history of their grandparents, an elderly neighbor, or someone older they are inspired by on Thanksgiving. Since 2015, thousands of high schools from across the country have "preserved more than 100,000 interviews, providing families with a priceless piece of personal history," according to the StoryCorps website.[25]

"Writing one's story not only boosts self-esteem and reduces stress and anxiety, it is also a powerful tool for a senior—or anyone—to visualize and create their future," says Hope Levy of There's Always Hope, a geriatric consulting firm, in an *AgingCare* article by Mike Brozda.[26]

When Ed Donakey teaches a class in ancestry at Brigham Young University, he asks his students to interview a grandparent or oldest living relative about their favorite memory in their life.

"I think it is so vital for us to know the stories of our elderly relatives and to pass them down through the generations," says Donakey, of FamilySearch International. He then tells the students to write down what their answer would be to the same question and save it so they can hand it down one day, too.

## Leading a Storied Life: Exploring the spaces where psychology, genealogy, and history converge

*"Telling your family stories helps you know who you are, connect with relatives, and preserve your family's legacy," says Steve Hanley, PhD, a clinical psychologist and amateur genealogist who works with and inspires people to learn about and tell their family stories. He draws heavily on his own story as a guide.*

*Working together through his Southfield, Michigan, area firm, The Psychogenealogist, Hanley and his team, including writing partner Geoffrey Bankowski, MFA, say they believe every person has a unique story to tell and they are dedicated to helping tell them. Their motto: "All lives are storied."[27]*

*"Any life, considered carefully, is incredibly complex," says Hanley. "What makes a person unique is his or her particular version of the struggle and enjoyment of life."*

*Hanley launched his own career as an amateur genealogist when he was in elementary school and he created a giant poster board collage of his ancestors. When he went to the University of Michigan, and later continued with his master's and doctorate degrees in psychology, he says he'd burn the midnight oil, staying up all night to "obsessively check the leaves of his family tree on Ancestry.com."*

*"I quickly learned it's a hit and miss game," he says. Three years ago, he had his own DNA tested.*

*His passion for helping people as a psychologist informs his approach to genealogy. He realized that people could better understand the impact of historical traumas across the generations. By going beyond just the facts of a family tree, one can better appreciate the larger family story that connects ancestors and descendants. Today, he applies these principles through a separate service, Your Storied Life: Unique Biographical Portraits (yourstoriedlife.com). He and his team work with people to really dig deeply into their stories without glossing over the parts they feel are shameful. In voicing their true stories, people can better understand what has truly shaped their lives and find empathy and compassion for themselves and the ancestors who came before them.*

From the front lines of his own family story, Hanley has pieced together the facts, the figures, and ultimately, the story of his father's side of the family, which carries the universal themes of many immigrants to America and the grit that it took to weather the hardships they endured.

Giving life to the theory that every picture tells a story, pictures of the family in their Irish homestead and their immigration journey to the Upper Peninsula of Michigan reveal a family strongly rooted in their Catholic faith—the foundation that most likely held them together through some tragic stories he uncovered. One of those stories was of a cousin on the Polish side of his family who, as a four-year-old, witnessed the murder of her nanny (a story discussed below in an excerpt from Hanley's blog), and another story was of a great-grandfather's suicide that was never talked about.

"When I approached my dad about [my great-grandfather's suicide], he said his dad always told him, 'We don't talk about it,'" says Hanley.

Memorabilia he discovered, including a pipe and a rosary, along with the photo album he discovered, make the stories that much more real.

It is his personal investment in understanding the connection between the lives of our ancestors and the impact they have in our daily living that Hanley says he brings to his clients.

### When DNA Breaks Open the Past

"One thing I've seen is that families that have an estrangement have a hunger to know the rest of the story, as crummy as it might be," he says. "With 7 to 10 percent of people doing their DNA getting surprises, people who have shut off parts of their pasts are suddenly having them broken open."

"Some of them are very curious about their past, and now with DNA the information is readily available," he says. "But we must be very careful here, because some people want to connect, and some do not want to open that Pandora's box. For those who do connect, it can be very healing. One person I know uncovered that her mom was Polish and had a hardship immigration story."

He adds, "The details in these family stories can be very therapeutic for people and can give them powerful insights about why they respond to certain things in a certain way. The stories help us better understand ourselves."

---

**Questions, Questions: What to ask yourself about your family story**

Hanley shares these tips for turning your research into your family story. Before you start, consider these questions:

- How have I become who I am?

- What did my ancestors think, feel, or worry about?

- Where did my family come from, and how does it shape me?

---

- What are my family's secrets? Why were they kept?

- Who in my past am I most like?

- How did historical events shape the personalities of my forebearers?

- Where can I learn about my family stories from historical documents?

- Where can I find those documents?

- What will I teach my children about our roots?

### Excerpts From the Psychogenealogist Blog

*One of Steve Hanley's goals is to help other people write their own family narratives by telling and sharing his own stories. Here, in an excerpt from his blog, the Psychogenealogist shares a poignant story on the two ancestors he discussed above. His from-the-soul depictions of his ancestors demonstrate how all of us can find meaning in the facts and intimate details we uncover when searching for our ancestors and trying to describe the life-changing impact they have had on us.*

## Quin Augustus Ryan

Quin Augustus Ryan (1898–1978) is the most famous relative I have written about to date in my #52Ancestors genealogy blogging challenge for 2018. He is my second cousin twice removed. Quin's maternal grandmother, Margaret Sullivan (1838–1925), was the older sister of my second great-grandmother, Ellen Sullivan Kelly/Hanley (1850–1938).

My third great-grandparents, Dennis Uonhi Green/Sullivan (about 1810–unknown) and Margaret Lowney (1815–1865), were Quin's great-grandparents. Dennis and Margaret lived in the village of Ballydonegan in County Cork, Ireland.

I connected with a Timothy Ryan on Ancestry.com. Timothy is a descendent of Quin's father's older brother, James W. Ryan (1863–1937). Tim provided a great deal of information about this family in his family tree.

## Quin A. Ryan (1898–1978)

Quin was the oldest of three siblings.

A search for "Quin Ryan" on newspapers.com, filtered for just Chicago, Illinois, gives close to five thousand articles! Though I haven't evaluated them all, it seems likely that most of these are related to the Quin profiled here.

Quin was a prolific and pioneering broadcasting legend in Chicago. Tim wrote a very nice write-up about his life and career. Here are a few of the highlights about Quin's life:

- His father was prominent Chicago-area attorney and judge Joseph Ryan (1870–1915).

- He attended Northwestern University in Evanston, Illinois.

- In college, Quin was active in writing and producing theatrical performances.

- Quin's first journalism job was at the *Chicago Tribune*.

- He became a pioneering voice in radio broadcasting.

- On April 14, 1925, Quin became the first ever to do a live radio broadcast of a baseball game (opening day for the Chicago Cubs at Wrigley Field, formerly known as Cubs Park).

- He was the first to broadcast a circus.

- He wrote a children's book, *Joey, the Littlest Clown*.

- Quin's live radio broadcasts included the Indianapolis 500 and both Republican and Democratic National Conventions in the 1930s.

- MANY other projects, endeavors, and accomplishments!

Quin married Roberta Nangle (1908–1944). They did not have any children. He died on October 7, 1978.

I love finding interesting stories like this in the lesser-known branches of my family tree. This one is a doozy! There is a lot more to learn and explore here. For instance, I am trying to find any archived audio files of Quin's broadcasts. I also found and purchased a copy of his children's book.

### *"Girl, 4, Is Witness to Murder": Delores Urkowski, (1930–1978)*

Dolores Urkowski is my second cousin once removed. She was the sole witness to the murder of her twenty-year-old nursemaid, Jennie Zablocki, in 1933. This is what I know of the story.

On the evening of December 6, 1933, Miss Jennie Zablocki was murdered at the home of my first cousin twice removed, Cecilia Pawlowski Urkowski (1908–1973) and her husband, Walter Urkowski (1901–1980). The only witness to the murder was their daughter, Dolores Urkowski. She was nearly four years old at the time.

Dolores's grandparents, John Pawlowski (1882–1920) and Paulina Grzeskowiak Pawlowski (1883–1941), were my second great-aunt and uncle. In fact, John was the brother of my great-grandfather, Adam Pawlowski (1879–1959), and Paulina was the sister of my great-grandmother, Marianna Grzeskowiak Pawlowski (1879–1941).

Two Pawlowski brothers married two Grzeskowiak sisters.

The murder took place at the Urkowski's home on 6338 Rugg Avenue in Detroit, Michigan. Miss Zablocki was employed as the nursemaid for the family taking care of Dolores and her younger brother, Donald. A third sibling, Walter Jr., was born later in 1937.

The primary suspect as reported by the *Free Press* was Elton Cebelak, Miss Zablocki's fiancé, whom she planned to marry on Valentine's Day the following year. The article states that Miss Zablocki was a distant relative of the Urkowskis, but I have no information about how she might be related.

The murder of Miss Zablocki made national news. There were several national news articles about the murder, including one published a week after the crime in the *Casper Star-Tribune* in Casper, Wyoming.

A follow-up article three days after the murder suggested that Mr. Cebelak had a strong alibi (he was at work) but that the police were still questioning him. This article also reported on some of four-year-old Dolores's account of the murder. She said that a man "struck Miss Jennie on the head with a stick."

From there the story took an interesting turn, as laid out in a *Detroit Free Press* article on December 11, 1933. A new suspect was charged, and Elton Cebelak was cleared!

This [article] reads like some good old-fashioned

1930s detective work. The new suspect, nineteen-year-old Ernest Di Oro, was arrested four days after the murder. He was identified by a companion, George Miotke, who apparently drove Di Oro to the Urkowski's home, unaware that Ernest had murder on his mind.

Here are the some of the more interesting highlights of the situation, as reported:

- The initial suspect, Mr. Cebelak, was strategically held at police headquarters in hopes of luring Mr. Di Oro, the real suspect, home thinking that he was cleared.

- Jealousy was the motive. However, it was Mr. Di Oro's jealousy in response to Miss Zablocki having "induced Di Oro's sweetheart to go out with another man."

- Mr. Di Oro was a milk wagon driver.

- Mr. Di Oro went into the Urkowski home empty-handed but came out wearing leather gloves, with bloodstained shirtsleeves, and carrying an accordion.

- You read that correctly: in addition to the murder, Di Oro STOLE AN ACCORDION that belonged to the Urkowskis.

The details of the murder at this point in the reporting were still a little murky. Some follow-up articles provided answers.

One from December 12, 1933, in addition to correcting the suspect's name to Di'Orio, details the man's confession, motive, and state of mind. Before his confession, in an effort to appear innocent, Di'Orio even attended the wake of his victim and donated fifty cents to a flower fund for her funeral!

One month later, on January 12, 1934, it was reported that Di'Orio was sentenced to twenty to forty years in Jackson Park for murder.

Miss Zablocki, obviously, is the primary victim in this case, and we shouldn't lose sight of that. That said, what drew me to this story was the fact that a four-year-old girl, Dolores Urkowski, witnessed such a horrific tragedy.

Dolores lived until 1978. Until I started researching my family tree, I had no knowledge that she existed or that this crime was part of my extended family. I need to research some more, but I don't believe Dolores ever married, and I don't think she had children. No one in my immediate family knew of this branch of the family tree.

## LEGACY LESSON

"What started out as a curiosity about my family heritage evolved into a deep and meaningful experience of connectedness. I feel close ties to ancestors I never met. My relationships with family, including newly discovered cousins, has deepened. Most surprising of all, perhaps, is the closeness I feel toward generations yet to come. Preserving and celebrating our shared family legacy has been time well and joyfully spent."

### Every Picture Tells a Story

"Sometimes the only 'story' people need to feel completed is a photo," says Diahan Southard, founder of Your DNA Guide. "They might not feel comfortable connecting to their adoptive mother or a long-lost sibling or pursuing that relationship. But a photo can be a very meaningful way for them to find closure. It helps them get a glimpse of their unknown family members. And they might see a common characteristic that is a huge benefit to them."

One interesting way to get a glimpse of what ancestors might have "really" looked like is to have treasured black-and-white family photos colorized.

**Pint-Size Pedigrees: Why family history connections are important for kids**

*Caroline Guntur, a professional genealogist and photo organizer, runs her own business, the Swedish Organizer LLC. She is committed to sharing her Swedish ancestry with her daughter.*

*"In our household, photos are a part of our daily lives, and as a parent I've always felt like it's important to share what I know about my heritage with my daughter," says Guntur. She assumed that Anica, a seventh grader, might be feeling bombarded by her stream of family history narratives. But much to her surprise, her daughter wrote this essay saying how much she appreciates their chats.*

*This essay, written by Anica Guntur, speaks volumes about the power of sharing our family stories with our children and the generations that follow us.*

### Why I Like Hearing Stories About My Family

I think it is important that we know about our past because then you will know who you are and what you came from. I know who I am because Mommy told me, and I think that is very important to know in life, especially when you are a kid, because sometimes you are not always so confident. Sometimes there are people in school who are not nice, but if you remember that you are nice then you don't have that problem. I think my family history is exciting because I get to learn what my ancestors' accomplishments were, and what their life was like, with pros and cons of what happened through life for them and how it was different than our life in modern times. I think it is exciting to sit down and learn about all the great people in our family's past.

I like photos, because when you look at them it reminds you of your past, and if you look at a photo of someone you love, for example, my family overseas, it's like you are right there with them and it goes into your memory often so you never forget each other.

I think other kids should know about their family history because first of all you will know what culture your family was from before you were in the country you are in now, or if your family was native to that country. Also, what your family has faced in the past and what has changed for you that was different for them. It's just so interesting to learn about it, and I think other kids would like it, too.

When Mommy and Daddy became American citizens, I was very proud because from that moment on I knew they would be just like me.

I am proud of one of my ancestors who was a great emperor. He was the first person in his time to let his daughters go to school, so he was very wise and smart, and when he was not first in line to be emperor, he didn't fight with his brother to get the throne. He just waited to see if he would ever get a chance and if not, oh well. And I think that is very great. He seems nice. He got a chance to be emperor when his older brother got sick and died. And so I am very proud that he was a good sport and very wise and chose decisions carefully. I wouldn't have known all of this if Mommy didn't tell me about it, and now I am proud. That's why you must tell your kids about these stories.

When I am older, I'm going to tell my kids about family history. For example, archery is a big hobby of mine, and I have learned that my ancestors liked it, too. So, when I'm older I might get my kids into archery and I might randomly say: Did you know your ancestor so and so did archery? This is better instead of sitting down and giving them a boring lecture.

Other parents can get their kids interested in family history by maybe taking a DNA test, which I think kids would like. It would show them what parts of the world they are from. I know I would like that. I already know mine. And after you know that you can find different types of food they would eat in those countries and try it. That is always interesting. And then at the end of the day you could sit down and read a book about your family's history to your child.

People must know about their family history. It is very important. And everyone, especially those people on TV, would be a lot nicer if they knew how their families lived before them.

## Traditions, Traditions: Exploring ancestry storytellers through our religious history

*Excerpt from Jaimie Eckert's blog[28]*

Typical ancestry searches today begin with the living subject and move backward in time. This is a relatively recent way of viewing genealogy. In Judeo-Christian history, genealogies began in the past and worked forward, predicated on one man (usually Adam or Abraham). Isaiah 41:4 says, "Who has performed and done it, calling the generations from the beginning? I, the Lord, and the first: and with the last I am He."[29] This gives the picture of divine foreknowledge being able to see every generation from the very beginning to the end of time.

The Hebrew word used for "calling" in this passage is *qara*, which means to proclaim or read. It is from the same Semitic root as the Arabic *qara'*, which is to read or recite. We get a picture of God unfurling a scroll and reading my genealogy from the time of Adam until now.

The queen of England has a genealogy purportedly starting all the way back with Adam. Jewish children in the time of Christ started reciting their ancestry from Abraham. The concept of reading an ancestry line *forward* instead of *backward* has very meaningful implications: it suggests that God had a specific purpose for Adam, for Abraham, for my specific ancestor, and it reads *forward* like a story of God's fulfillment of promise.

I like to memorize portions of the Bible, and this

December I memorized Matthew chapters one and two, on the birth of Christ. The majority of chapter one is the genealogy of Jesus, and it seemed daunting at first to get past all the "begats." What I discovered is that the ancestry portion proved to be much easier to commit to memory than the narrative portions. Why? Originally, Scripture was received in a context of orality rather than literacy. The ancient Hebrews used mnemonic devices and memory-aiding structures to help listeners not just understand, but also memorize the text.

The ancestry of Jesus in Matthew 1 is compiled in triplets of three names each and divided into three sections of fourteen generations (each set of fourteen having one triplet with an abnormal number to total fourteen). Although I've read Matthew 1 many times, I didn't discern the underlying memory aids until I set about to commit it to memory for myself. But it makes sense. Jewish children would have been memorizing their ancestry from very early ages—starting at Abraham and moving forward, reminding them that God had made a promise to their patriarch and still had a purpose for their lives.

Muslims are interested in genealogy in varying degrees. In a broad sense, Shi'ites tend to be more interested than Sunnis because of historic differences between the two sects. Sunnis and Shi'ites split from each other over an argument about who would succeed the prophet Muhammad—an elected leader, or a biological descendant?

Shi'ites took the latter opinion and have attempted to continue tracing the prophet's lineage throughout the years. It is not always straightforward, since Muhammad had no male descendants, but it is nevertheless a cause for robust scholarly discussion.

In Islam, blood relations are very important, and this is one of the reasons why adoption is not permitted in traditional Islamic contexts. Although Islam strongly encourages its adherents to care for orphans, children raised by others will always retain the name of their biological family and will not automatically inherit from "adoptive" families who care for them (a relationship sometimes called *kafala*, which is a term that can also mean "sponsorship"). In ancestry searches, this would theoretically remove certain challenges since you know you never need to deal with name changes due to adoption. In general, Islam retains a more robust view of blood ties than Judeo-Christian traditions, and this may in part be because blood relations fundamentally inform *mahram* regulations regarding whom you may marry and with whom you must observe modesty codes.

## CHAPTER TEN

## NAVIGATING YOUR ANCESTRY JOURNEY:

### Resources to help you unlock your past and uncover your true beginnings

Now, more than ever, Americans are looking to their ancestors to guide them as to what they may become during an age characterized by mobility and a rootlessness.

Despite the challenging information we might discover, we want to know the truth. When these are stories we can hardly imagine, they nonetheless remain stories we yearn to embrace. Certainly, narrative is the biggest part of history, and everybody has a story. But each person's journey to the past is different.

For many, this process is opening a can of worms:

questions about heritage, ethnicity, race, culture, and privacy. Many ask the question, "What next?" after seeing their ethnicity breakdown and the hundreds of cousin matches that came with their DNA results.

We look to our ancestors for enlightenment, for spiritual inspiration to help guide us where we are going. At a crucial time in our world, when we are becoming increasingly mobile, we yearn to travel in our minds and hearts into the lives of our ancestors and also physically to their places of origin, because to have come from a specific place, no matter how long ago, is to be connected to a much more meaningful self-narrative.

The timing for us to tell our stories is critical. Many first-, second-, and third-generation immigrants are more closely connected to their family's roots, but younger generations are farther removed from the ancestral ties. They need the tools and the inspiration to know that the journey to the past can be closer than it seems. Take the six degrees of separation concept. The person next to you at a German Oktoberfest may very well have relatives who hail from the same town as your ancestors. With a little sleuthing, we can uncover amazing tales about those who came before us.

For younger generations, they might have grandparents or great-grandparents who have faded black-and-white photos depicting them as one of the "huddled masses." The memories—the lives, talents, hopes, and dreams—of those who came here on the boats from Europe or migrated to distant cities are disappearing as the generation of immigrants from the 1920s is fast slipping away.

During the interviews for this book, it became clear that we are treading new ground about what it means to research our family heritage as we find ourselves on the cusp of what is predicted to be an explosion in DNA testing.

So, what is the blueprint for telling our family stories now? How do we respond to, record, and answer the questions that are popping up everywhere?

- What do I do if I just learned my dad is not my real dad? I feel blindsided.

- Should I call the woman I have found out is my biological mother?

- Should I keep this a secret or tell my other family members?

- What do I do when I've hit a roadblock?

Experts say the best antidote to discovering secrets that seem painful in our family stories is to find a new way to tell and reframe our stories. By doing so, we can heal the wounds for our entire lineage—wounds that have been holding those who came before us captive for years.

One thing that is clear is the importance of storytelling in helping searchers find meaning that allows them to move ahead with these shocking revelations.

Here, I have tried to compile resources and tips from the experts I have interviewed that can help give you support for unexpected DNA matches and stories in your family history.

"People don't have to collect, scan, and memorialize everything," says Caroline Guntur, genealogist. "It's better to have one solid curated story with ten photos rather than thousands of photos in a digital mess."

She adds, "It's okay to be selective and delete what you don't need. Sometimes it feels overwhelming, and that keeps people from even starting. I think it's important to remember why you're doing it and just start instead of getting paralyzed because there's too much."

### How to Prepare for and Cope With DNA Discoveries

In her role helping people cope with stress, Anita DeLongis, PhD, has been inundated with requests in the past year to help people who are struggling with the angst resulting from shocking DNA discoveries. In Chapter One, she describes the study she is conducting on the impact of revealing information obtained from DNA tests and ancestry searches with her team at the University of British Columbia's Centre for Health and Coping Studies in Vancouver.

Here, DeLongis offers some tips for coping when the results have been surprising.

**Do some anticipatory preparation.** Before you decide to do a DNA test and/or start hunting and pecking for family history on the internet, ask yourself:

- What is my goal?

- What are the possibilities that can happen?

- How would my family react with surprising results?

**Enroll a confidant and support person.** Once you have your results, pause for a moment before you race into telling everyone about them. Find someone in your circle whom you can trust with the information and create a plan for how you're going to approach sharing the news. That person can be a trusted friend or even a therapist, but it should be someone outside the circle of those who will be impacted. Ask yourself how this information will affect others.

**Consult a professional.** Before you share what could be shattering information for others, consult a DNA specialist to help you interpret the findings. Don't assume you are certain about the information.

**Positively reframe the information.** Approach the findings from a protective stance and an open mind. Consider putting the information in its historical context. Why would your biological mom have to lie about giving birth? Were your family members trying to protect you by lying about who your real family is? Remember there was a time when someone would have been called a "bastard" if he or she was illegitimate.

**Look for the silver lining.** Consider all the different family narratives that could have happened and try to find the positive.

### A Blueprint for Handling the Initial Outreach to Newly Found Relatives

Currently the road map for navigating biological relationships is not fully developed. Genetic genealogist Diahan Southard, whose mother was adopted, suggests blazing a trail with kindness.

She knows firsthand there's no real blueprint for the best way to handle a first contact—receiving or sending the surprising news that you are family members.

But in the meantime, Southard, founder of Your DNA Guide, shares these insights.

**Pause**

Don't react immediately. Find a confidant and talk through what this will mean for your family and the family you are reaching out to. It's easy when you discover the information to go with the thrill of the moment and act impulsively. Don't.

### Take the lead on the communication

Take the communication seriously. Examine carefully if you should tell or not tell. Understand that some people might not want to know, and explore what the impact might be. Be thoughtful. Be intentional.

### Be responsive

Always respond to any communication, even if you are declining someone's pursuit of more information. Try to understand their situation and treat them like you would want to be treated if you were delivering the news.

### Have patience and empathy

It can be frustrating when you do not receive a response. Some people might not want to make the connection. Recognize that everyone has the right to engage in the relationship or to choose not to be involved.

### Communicate via email

Proceed with caution and make sure the communication is private. Keep the message simple, with a short introduction in a friendly, personal tone, and end with a specific, easy-to-answer question.

## *Answering Communication from DNA Matches*

### Appreciate their interest
This person has taken the time to reach out to you. They're essentially cold-calling you because you're genetic family.

### Respond promptly
When someone reaches out, they're paying attention, so take advantage of their enthusiasm. A quick response from you puts your match to work for you.

### Identify what you really want

- Have a goal of your own.

- Be proactive about your own goals and priorities.

### Organize the information
Put that inquiry in a folder that will remind you to look at it later when you have a few more minutes.

### The Dos and Don'ts of Family History Research

Genealogist Mary Hall offers these tips excerpted from her blog, Heritageandvino.com.

- **Do** keep track of your sources—one day you'll wonder how you know your third great-grandmother's baptismal date, but you won't know how to find it to check who her sponsors were or what church it happened in.

- **Don't** copy someone else's work, especially if they haven't documented any sources. Tons of erroneous info is passed along the internet— don't fall for it or pass it along!

- **Do** double-check someone else's sources— they might not have copied all the relevant or interesting information, or they might have read a transcription wrong.

- **Don't** trust the transcription—names are spelled wrong all the time and, again, while a transcription might list a name, a date, and parents' names, the actual record might include a parent's occupation, sponsors' names, an address, and other useful informa- tion. (A transcription is a secondary source, someone else's copy or interpretation of the

243

information, unlike a document that is a primary source, such as a birth certificate or marriage license.)

- **Do** use primary and secondary sources whenever possible—if you have a transcription, try to find the actual record. If you're using a compiled family history, try to find the actual sources they used. The more times information is copied, the farther away you get from the actual primary and secondary information, the more chance there is for errors to be made and passed along.

- **Don't** keep your discoveries to yourself—we trace our family trees for our families, so if a distant cousin reaches out to you and needs help, share your work with them. You might never have met them, and they might be your fourth cousin five times removed, but, hey, they're family!

- **Do** take breaks—family history research is mentally and emotionally exhausting, and especially when we're constantly hitting our heads against a stone wall, it can be easy to want to call it quits. Sometimes it can be good to walk away for a month, a week, a day, an hour, clear our minds, and come back to our research with fresh eyes.

- **Don't** get discouraged—genealogy is hard. Records aren't available. Handwriting is impossible to read. Government agencies take forever to send you the marriage record you need to find out your great-grandfather's parents and place of birth. Some of these problems will never be resolved, such as records that have been destroyed, so there's no use in getting upset over that; everything else might require just a little patience.

- **Do** learn a new language—I don't mean become fluent in Russian or Dutch. I mean, whatever countries your family originated in, whatever languages they spoke, learn to recognize the important genealogical words in that language—born, baptized, married, died, buried, parents, godparents, from (where they lived), etc. You might not be able to glean all the details of an entry, but if you're looking through an eighteenth-century German church book, I guarantee it'll be in German, and if you know these key words, they will absolutely begin to pop out at you and you will be able to take advantage of what might be an invaluable family record.

- **Do** keep a cheat sheet handy—keep a list of those words nearby. And if the region used a different alphabet than you're used to, keep a cheat sheet of those letters and symbols, too. I do a ton of German heritage research, and they used four different standard alphabets depending on the place and time period; my cheat sheet is always right next to my computer.

- **Don't** forget about history—it's important to put our families into context. Maybe the entirety of world history is too overwhelming to cover and not quite relevant to our families, but if you can put an ancestor in a time and place, finding out more about what was going on in that time and place can help you understand things such as what a person's occupation might have been, why a family might have emigrated out of an area, what a person's religion might have been, whether or not someone might have served in the military, why a mother had five children all die young in a two-year period, and so on.

- **Do** look at other branches of your family

tree—it's tempting to just focus on our direct lines, but following sibling branches is extremely helpful in connecting to cousins, close or distant, which in and of itself can be rewarding, but which also sometimes yields a wealth of family history information that you might never have known about but that got passed down to them. Also, if you've been hitting your head against walls, sometimes the information you're looking for—a birthplace, a parent's name—that is missing on your direct ancestor's documents can be found on the documents for one of their siblings, effectively opening up that dead end.

- **Don't** think you're ever finished—a genealogist's work is never done, nor do we want it to be! I have been doing this on and off for twenty-five years, and while there have been lulls in discovery, I am still uncovering people and places and information that are new and exciting and opening up branches of my tree I never dreamed about.

- **Do** get out there as soon as you're done reading this and start or keep digging!

- **Don't** wait!

## Looking for Your Ancestors in All the Wrong Places: Creative ways to continue your search

*Blog excerpt by Carol DiPirro-Stipkovits, president of the Niagara County Genealogical Society and a member of the National Genealogical Society. She has been doing family research for more than fifteen years and blogs at noellasdaughter.com.*

### Clues Written in Stone: Headstones With Stories to Tell

Where your ancestors are buried can tell you so much about them and their lives. Is your ancestor buried in a church cemetery? They were likely involved in a faith community. If they were buried in a family cemetery, consider that it was very likely part of family land at one time.

Grave markers are a direct connection to our ancestors' lives and may have clues, literally, written in stone. There is so much more to cemetery research than just the names and dates on the gravestones. Look around to see who is buried near your ancestors. It's likely you will find connections, which may lead you to break down a brick wall within your family history.

Here are some basics:

Check death certificates, obituaries, and funeral home records to identify the cemetery where your ancestor is buried. Also, look at close relatives of your ancestor. If you've located where their sibling is buried, reach out to the cemetery office and inquire about others with the same surname.

Once you have the name of the cemetery, you'll need to locate it. Findagrave.com and Billiongraves.com allow users to search for cemeteries around the world. Billiongraves.com allows users to collect photos of headstones and upload them to their site by using a phone camera app. Once uploaded, the photo is tagged with the GPS location and becomes available to all users. (I located an ancestor's tombstone in Italy through this site!)

Now that you're ready to explore, organization is key.

With that said, repeat after me…it is not weird to have a graveyard kit. Mine includes: a plastic pail, scissors and a trowel, wet wipes, garden gloves and disposable latex gloves, insect repellent, cemetery map with grave location marked, masking tape, rags, flashlight, cheap aluminum foil, whisk broom, notebook and pen, plastic grocery bags (I like clean knees, too!), drinking water, water jug that can easily be refilled, sunscreen, soft toothbrush, and old shoes. Keep your phone charged and handy, not only for taking photos but for your safety. With safety in mind, it's best to have a partner with you.

By studying tombstones, we can discover facts about an ancestor such as hobbies, occupations, organizations, family members' names, and military service. They may also include cause of death. While visiting a Boston cemetery, I located a tombstone showing a man pinned under the wheels of a cart pulled by running horses, a tragic event memorialized for eternity. Even the nearby plantings may be symbolic; oak trees represent strength, while weeping willows are the symbolic tree of sadness. Looking closely, you may see symbols that held greater meaning in a time when many people didn't know how to read. Photograph the stones and notice the carvings, initials, and symbols. A Google search can easily decipher these. I consider cemeteries sacred ground where tombstones stand as monuments to an ancestor's life, filled with rich genealogical details just waiting to be unearthed.

### What to Do When You Hit a Brick Wall

*Hazel Thornton, founder of Organized for Life (www. org4life.com), offers these tips for when you are stuck on your ancestry search journey.*

*Look at these reasons why you can't find your ancestor.*

Some of them may lead you to a breakthrough...or, if nothing else, give you some insight:

1.  The record doesn't exist—some never existed to begin with (like my grandma's birth certificate), or they are lost to time or natural disaster (like the 1890 census, which was largely lost to fire). Or maybe you are looking

for civil records, when your ancestor was a Quaker and you really need to be looking at the Society of Friends' meeting minutes.

2. The record exists, but it's not online (yet?). There is still great value in visiting libraries, cemeteries, and other repositories of genealogical records! The one you need may be literally sitting on a dusty shelf in the basement of the city hall where your ancestor lived. (Ask me how I know.)

3. The record is online, but it's not indexed (yet?). Thus, it is not easily searched for and found. Volunteering on Family Search to help index records won't help you find your ancestor, but it will help other people find theirs.

4. You haven't yet recognized or accepted that the "wrong" surname spelling might, indeed, be your ancestor! (There are enough reasons why this happens to fill an entire new blog post.)

5. There's a typo (or other error) in the index, transcript, abstract, database, family history, or other derivative source. Check the original record if you can.

6. Terrible handwriting, and misreading of perfectly good old-fashioned script, causes errors at every level (original document, transcription, indexing).

7. The record or index used initials only.

8. Your ancestor truncated, Anglicized, or completely changed his name . . . but not because they forced him to at Ellis Island, because that's a myth.

9. Your ancestor went by a middle name, or a nickname, and all you know is their first name—or vice versa—and they switched back and forth over time. Or whoever reported or recorded the data did.

10. It's an original record, but it's wrong. Who supplied the information? Would they necessarily know? How close in time to the event—birth, marriage, death, etc.—was the record made?

11. The census taker interviewed whoever answered the door (whether or not they were the best person to ask); he wrote down what he heard (without concern for spelling) or he was tired or confused (or drunk!) and skipped that street altogether.

12. The census says your ancestor was [a different race] because the census taker reported what he saw or was told. (I have an ancestor who was reported as black, white, and mulatto in different census years . . . and she was a slave owner, too . . . the plot thickens.)

13. People were living together in combinations you didn't expect. (Who are these people with whom my great-grandfather is living as a child? Oh. . . . I see . . . his father died, and his mother remarried, that's all.)

14. On the Ancestry search form, you have too much data in the search fields. Delete some of it. Or add an educated guess (e.g., a good estimate for a parent's birth year is twenty years before first known child's birth year) to see if it helps generate results.

15. You are looking for records of your female ancestor's birth, but you are using her married surname. (In Ancestry, if you have her in your tree by her married name, delete the married name so Ancestry doesn't think it's her maiden name. Better to leave it blank if you don't know her maiden name yet.)

16. Your ancestor may have been married multiple times. Which surname, or spouse, or children still living at home, would apply to the time frame you are seeking to know more about?

17. Your ancestor may have moved. Or, in one case, it turned out mine didn't move at all, but the county lines kept changing around him from one census decade to the next.

18. You are looking only for vital records (birth, marriage, death). Try something new like land records, pension files, and newspaper articles.

19. You are focused only your ancestor. Try broadening the search a little, learning more about their FAN club (family/friends, associates, and neighbors).

20. You have made an error somewhere and are now off climbing someone else's tree.

## 20 Greatest Questions to Ask Living Relatives

*Excerpted with permission from FamilySearch's "52 Questions in 52 Weeks" project and blog.*[30]

1. What is your full name? Why did your parents give you that name?

2. When and where were you born? Describe your home, your neighborhood, and the town you grew up in.

3. Tell me about your father (his name, birth date, birthplace, parents, and so on). Share some memories you have of your father.

4. Tell me about your mother (her name, birth date, birthplace, parents, and so on). Share some memories you have of your mother.

5. What kind of work did your parents do (farmer, salesman, manager, seamstress, nurse, stay-at-home mom, professional, laborer, and so on)?

6. Have any of your family members died? If so, what did they die from? What do you remember of their death, and what were the circumstances of their death?

7. What kind of hardships or tragedies did your family experience while you were growing up?

8. Are there any unusual genetic traits that run in your family line?

9. What are the names of your brothers and sisters? Describe things that stand out in your mind about each of your siblings.

10. What were some of the family traditions that you remember?

11. Did your family have special ways of celebrating specific holidays?

12. Share a few memories of your grandparents.

13. Did your grandparents live close by? If so, how much were they involved in your life? If they lived far away, did you ever travel to visit them? What was that like?

14. Who were your aunts and uncles? Do you have any aunts or uncles who really stand out in your mind? Write something about them (names, personalities, events that you remember doing with them, and so on).

15. Where did you go to school? What was school like for you?

16. What were your favorite subjects in school? Why?

17. What subjects did you like the least? Why?

18. Who were some of your friends in school? What were they like? What are they doing today?

19. If you went on to get a college or vocational education, what school did you go to? What did you study? What memories do you have of those years?

20. What do you see as your strengths?

## ACKNOWLEDGMENTS

On the day my mother asked me to find her father who had disappeared when she was just two, I set off on a journey armed with less than a handful of clues—his first and last name, a faded photo of a thirty-ish looking man in a uniform, and a membership to Ancestry.com.

None of this would have been possible without the generosity of friends, family, and strangers who shared their time, care, and support to help me dig for information and break through many of the roadblocks I faced in the search. I'd like to thank the genealogy group at the Irish American Heritage Center in Chicago who teamed up on many occasions to pour through documents and help me start building the story of my ancestry. I'd especially like to thank Virginia Gibbons and the memoir-writing group who patiently listened to my readings about my grandma and grandfather and encouraged me to keep on writing and searching. Who would have known that over the course of three years, I would discover that one of the members would turn out to be my cousin.

I have to thank Marion Collins, Paddy and John Barry, and the members of the book club at Paddy's on the Square/The Irish Boutique in Long Grove, Illinois who have become friends and family over the past ten years. I am forever grateful for the support of the "Marys, Marions, and Mariannes" in our club who helped me find and discover the places both my grandmother and grandfather lived on our several treks to Ireland.

Thank you to my brother, Paul, sister Sheila, and sister-in-law Beth who so enthusiastically responded to my emails and texts in response of each new discovery. And, thank you to my children, Caitlin, Thomas, and Emily, nephew Michael, and grandchildren, Rylee, Tommy, and Keira, who will carry on the legacy of our ancestors for generations to come.

I appreciate the guidance and support of Hannah Bennett and the editors at Start Publishing for their guidance and patience in bringing this book to life.

A heartfelt thanks to all the people who generously shared themselves and their stories with me, and to the ancestors who came before them who brought to life these stories.

Lastly, thank you to my grandmother, for inspiring me to wonder and search, to find meaning in and complete the stories of those from where we come.

## ABOUT THE AUTHOR

Mary Beth Sammons is an award-winning journalist and author of twelve books, including *Living Life as a Thank You: The Transformative Power of Daily Gratitude* and *The Grateful Life: The Secret to Happiness, and the Science of Contentment* (Viva Editions). She is a cause-related communications consultant for numerous nonprofits and health-care organizations, including Rush University Medical Center, Cristo Rey Network schools, and more. She lives in Chicago's suburbs. Mary Beth's experience has run the gamut. She has been the bureau chief for *Crain's Chicago Business*, a features contributor for the *Chicago Tribune*, *Family Circle*, *Psychology Today* and a daily news reporter for *The Daily Herald* and *AOL News*. She is an active volunteer at the Irish American Heritage Center in Chicago and on the board of the Irish Books Arts and Music (IBAM) annual cultural event.

## ENDNOTES

1   Frank Delaney, *Shannon* (Random House Trade Paper-
    backs, 2010), 3.

2   "Customer Care page," 23andMe.com, "What unex-
    pected things might I learn from 23andMe?" 2020, https://
    customercare.23andme.com/hc/en-us/articles/202907980.

3   Pat Mertz Esswein, "Discover Your Roots With DNA Test-
    ing," *Kiplinger's Personal Finance*, October 3, 2019, https://
    www.kiplinger.com/article/spending/T065-C000-S002-
    discover-your-roots-with-dna-testing.html.

4   Global Market Insights, "Direct-to-Consumer Genetic Test-
    ing Market to Hit $2.5 Bn by 2024: Global Market Insights,

Inc.," PRNewsWire.com, December 11, 2018, https://www. prnewswire.com/news-releases/direct-to-consumer-genetic- testing-market-to-hit-2-5-bn-by-2024-global-market- insights-inc--830436085.html.

5    Antonio Regalado, "More than 26 million people have taken an at-home ancestry test," *MIT Technology Review*, February 11, 2019, https://www.technologyreview. com/2019/02/11/103446/more-than-26-million-people- have-taken-an-at-home-ancestry-test/.

6    Nikki Graf, "Mail-in DNA test results bring surprises about family history for many users," Pew Research Cen- ter, August 6, 2019, https://www.pewresearch.org/fact- tank/2019/08/06/mail-in-dna-test-results-bring-surprises- about-family-history-for-many-users/.

7    Ibid.

8    Wendy Bottero, "Practising family history: 'iden- tity' as a category of social practice," Wiley Online Library, July 14, 2015, https://onlinelibrary.wiley.com/doi/ abs/10.1111/1468-4446.12133.

9    Amy Dockser Marcus, "The Family Secret Uncovered by a Writer's DNA Test," *Wall Street Journal*, January 4, 2019, https://www.wsj.com/articles/the-family-secret-uncovered- by-a-writers-dna-test-11546626551.

10   "About Us," Vanderbilt University, 2020, https://wp0. vanderbilt.edu/fndp/about-us/.

11   Ree Hines, "Carson Daly Reveals the Sweet Story Behind Baby Goldie's Name," NECN, March 27, 2020, https:// www.necn.com/entertainment/entertainment-news/car-

son-daly-reveals-the-sweet-story-bsehind-baby-goldies-name/2252160/.

12 Kristine Celorio, "Irish I Were Mexican" (blog), http://irishiweremexican.com/at-post.html.

13 Sam Stanton and Darrell Smith, "How detectives collected DNA samples from the East Area Rapist suspect," June 1, 2018, https://www.sacbee.com/latest-news/article212334279.html.

14 Laurel Wamsley, "After Arrest of Suspected Golden State Killer, Details of His Life Emerge," NPR/WBEZ.95.1, April 26, 2018, https://www.npr.org/sections/thetwo-way/2018/04/26/606060349/after-arrest-of-suspected-golden-state-killer-details-of-his-life-emerge.

15 Ryan Osborne, "A Colorado teen's murder was cold for 38 years. This cracked the case in weeks," ABC/TheDenverChannel.com, September 16, 2019, https://www.thedenverchannel.com/news/local-news/a-colorado-teens-murder-was-cold-for-38-years-this-cracked-the-case-in-weeks.

16 "How Two Lives Met in Death," *Newsweek,* April 14, 2002, https://www.newsweek.com/how-two-lives-met-death-143283.

17 Valerie Wingfield, "The General Slocum Disaster of June 15, 1904," New York Public Library, June 13, 2011, https://www.nypl.org/blog/2011/06/13/great-slocum-disaster-june-15-1904.

18 Ibid.

19 Sarah Jean Green, "'A wrong had finally been righted': Tribes bury remains of ancient ancestor known as

Kennewick Man," *Seattle Times,* February 19, 2017,https://www.seattletimes.com/seattle-news/tribes-bury-remains-of-ancient-ancestor-also-called-kennewick-man/.

20  Emma Sparks, "Top Travel Trends for 2019," *Lonely Planet*, October 22, 2018, https://www.lonelyplanet.com/articles/top-travel-trends-for-2019.

21  Michael O'Donnell, "Heritage Travel on the Rise with Airbnb and 23andMe," New York PR Web, June 13, 2019, https://www.prweb.com/releases/heritage_travel_on_the_rise_with_airbnb_and_23andme/prweb16379006.htm.

22  Christian Jarrett, "The Benefits of Thinking About Our Ancestors," The British Psychological Society's *Research Digest*, December 20, 2010, https://digest.bps.org.uk/2010/12/20/the-benefits-of-thinking-about-our-ancestors/.

23  Peter Fischer, Anne Sauer, Claudia Vogrincic, and Silke Weisweiler, "The ancestor effect: Thinking about our genetic origin enhances intellectual performance," *European Journal of Social Psychology* 41, no. 1 (February 2011): 11-16, https://doi.org/10.1002/ejsp.778.

24  Christian Jarrett, "The Benefits of Thinking About Our Ancestors," The British Psychological Society's *Research Digest*, December 20, 2010, https://digest.bps.org.uk/2010/12/20/the-benefits-of-thinking-about-our-ancestors/.

25  "The Great Thanksgiving Listen," StoryCorps, 2003-2020, https://storycorps.org/participate/the-great-thanksgiving-listen/.

26  Mike Brozda, "The Power of Telling Family Stories,"

*AgingCare*, 2020, https://www.agingcare.com/articles/sharing-family-history-95687.htm.

27 Steven J. Hanley, "Your Storied Life," *The Psychogenealogist* (blog), https://www.psychogenealogist.com/your-storied-life/.

28 Jaimie Eckert, JaimieEckert.com (blog), 2020, https://jaimieeckert.com/.

29 Isaiah 41:4 (Holman Christian Standard Bible).

30 Steve Anderson, "52 Questions in 52 Weeks: Writing about Your Life Has Never Been Easier," FamilySearch blog, September 28, 2015, https://www.familysearch.org/blog/en/52-questions-52-weeks-writing-life-easier/.